HANDBOOK OF
BAKERY AND CONFECTIONERY

Bakery products, due to great nutrient value and affordability, are an element of huge consumption. Due to the rapidly increasing population, the rising foreign influence, the emergence of a working population and the changing eating habits of people, they have gained popularity among people, causing significantly to the growth trajectory of the bakery industry. The Handbook of Bakery and Confectionery delineates a theoretical and practical knowledge on bakery and confectionery.

Chapter 1-21: This part deals with basic concepts in baking and includes chapters on all bakery ingredients and their functions, bakery products in the baking industry.

Chapter 22-23: This section provides an affluent information about production of various chocolates and toffees.

Dr. S.M.D. Mathuravalli, M.Sc., M.Phil., Ph.D., is currently working as an Associate Professor in Holy Cross Home Science College (affiliated to Manonmaniam Sundaranar University, Tirunelveli), Thoothukudi. She has 25 years of experience in teaching, research and extension programmes. She has guided many postgraduates and M.Phil., Scholars in the field of food and nutrition. During her extended career, she has actively partaken and presented papers in national and international workshops, conferences and seminars on Food Science, Nutrition and Dietetics and won several prizes and awards including the best paper and best presenter.

HANDBOOK OF
BAKERY AND CONFECTIONERY

By

S.M.D. Mathuravalli

Department of Food Science and Nutrition
Holy Cross Home Science College
52, New Colony
Thoothukudi - 628003

CRC Press
Taylor & Francis Group
Boca Raton London New York

CRC Press is an imprint of the
Taylor & Francis Group, an **informa** business

NARENDRA PUBLISHING HOUSE
DELHI (INDIA)

First published 2022
by CRC Press
2 Park Square, Milton Park, Abingdon, Oxon, OX14 4RN

and by CRC Press
6000 Broken Sound Parkway NW, Suite 300, Boca Raton, FL 33487-2742

© 2022 Narendra Publishing House

CRC Press is an imprint of Informa UK Limited

The right of S.M.D. Mathuravalli to be identified as author of this work has been asserted in accordance with sections 77 and 78 of the Copyright, Designs and Patents Act 1988.

Print edition not for sale in South Asia (India, Sri Lanka, Nepal, Bangladesh, Pakistan or Bhutan).

British Library Cataloguing-in-Publication Data
A catalogue record for this book is available from the British Library

Library of Congress Cataloging-in-Publication Data
A catalog record has been requested

ISBN: 978-1-032-15126-7 (hbk)
ISBN: 978-1-003-24263-5 (ebk)

DOI: 10.1201/9781003242635

Printed in the United Kingdom
by Henry Ling Limited

Contents

Preface

In an increasingly globalized world and the fluctuating paradigm of urbanized living, the demand for bakery and confectionery has amplified manifold the world over. During my 25 years of industrial and teaching tenure in the field of Food Science & Nutrition, I was well accustomed with basic necessities of students in the Bakery and Confectionery field. That is hindsight from my earlier days and this has made me to reminiscence my initial days as a pupil, when I was in need of a book on bakery in a simple language. This mandate has facilitated me to emerge with a book which would really comfort even the moderate learner by giving them adequate info in an accessible style. The language used in this book is very simple with lot of pictorial illustration. One decade ago, I have prepared a Students Handbook on Bakery which was meant to channelize my students. That was my first landmark and this book is my second landmark in my career. I felt, if I launch this book, it will be helpful for Catering, Hotel Management, Food Science, Nutrition and Home Science students. Here, I have rationalized more information and metaphors to provide the learners about the complete product which will be very helpful to them. I hope this text will serve a worthwhile resource in this subject. I would like to thank Rev.Dr.Sr. Mary Gilda, Secretary & Principal of Holy Cross Home Science College, Thoothukudi for their support in preparing the book. I am very much grateful to M/S Jaya Publishers, Delhi in the timely publication of this book. I would like to thank my husband, Dr. N. V. Sujathkumar and my sons Mr. M. Sakthi Ragavendran and Mr. S. M. Ramesh Krishna and My daughter, Ms. M. S. Kirthika for strong support and encouraging me in completing this book. I wish to thank my colleagues, students and friends for the encouragement. Finally, I am thankful to the Almighty for His abundance of grace and blessings to bring out this book. Before I come to a close, I share with you my joy and pleasure of releasing this book. Remarks and recommendations are most welcome for further progress of the book.

S.M.D. Mathuravalli

CHAPTER 1

INTRODUCTION TO BAKERY

HISTORY OF BAKING

The organized production of wheat by the Egyptians is considered by most historians to be the beginning of the breads produced today. Many centuries after the Egyptians (about 400 B.C), the Greeks were preparing more than fifty kinds of bread, all baked in closed ovens. The Romans united the Greek and Egyptian developments in bread making with their own developments to start producing the bread in large scale. During the reigns of the emperors Augustus and Julius Caesar (100 to 44 B.C), public bakeshops were established in the cities of Roman Empire. Pastries of various kinds were traded to spectators during the games in the colosseum.

While Roman civilization spread throughout Europe, the Middle East, and the North Africa, the new profession of baking was born. Baking knowledge grew through experimentation and the influx of information from new conquered territories.

However, with the slow degeneration and collapse of the Roman Empire, the new baking industry also collapsed. Knowledge, the true legacy of Rome, was preserved in monasteries, and during the Dark Age the temporarily lost art of baking was practiced mainly by monks who kept their baking knowledge as well-guarded secret for many years. At the beginning of the thirteenth century, Philip II of France granted bakers the right to build their own ovens. This movement by Philip against the power of the nobles and the church resulted in the incorporation of the Patissier Dublayers of Paris in 1270. Those were pastry and bread specialists, and, with an industry incorporating both professional baking was once more firmly deep-rooted.

The industry continued with only slight changes until the discovery of America and the influx of new ingredients, particularly sugar and cocoa. In 1675, the baking art was given another boost when a Sicilian pastry cook named Procopio went to Paris and opened the first ice cream parlor. This success gave rise to Dublayers who roamed the streets of Paris selling galattes and sweet breads. The distinction between pastry cook and baker became more clear in the early eighteenth century. Bakers and pastry makers separated generally because of arguments about proper oven temperatures (bread requires a much stronger heat than delicate pastries). In 1790, the first school of baking opened its doors in Paris. The French Revolution unbound servant-chefs of French aristocrats. These culinary masters could now offer their knowledge and talents to the public.

DEFINITIONS
Bakery

- A Bakery is an establishment which produces or/and sells bread, pastries, cakes, biscuits, cookies etc.

- A Bakery (or baker's shop) is an establishment which produces and sells flour-based food baked in an oven such as bread, cakes, pastries and pies.

Baking

- Baking is the cooking of food by dry heat in an oven in which the action of the dry convection heat is modified by steam. The dry heat of baking changes the form of starches in the food and causes its outer surfaces to brown, giving it an attractive appearance and taste. The browning is produced by caramelization of sugars.

- Baking is a method of preparing food that uses dry heat, normally in an oven, but can also be done in hot ashes, or on hot stones. The most common baked item is bread, but various types of foods are baked. Heat is progressively transferred from the surface of cakes, cookies, and breads to their center.

- When baking, steam rises from the water content of the food; this steam combines with the dry heat of the oven to cook the food, e.g. Cakes, pastry, baked jacket potatoes.

Baker

A person who prepares baked goods as a profession is called a **BAKER**.

Characteristics of a Good Baker

- Passion and good leadership.
- Good numerical skills.
- Creativity.
- Able to work under pressure.
- Good organizational skills.
- Awareness of safety and hygiene rules.
- Reasonable level of physical fitness.
- Ability to work in a team.

Changes during baking

The reactions in this kind of dough occur more rapidly (when baking). Both Baking Powder and Baking Soda react with chemicals in the dough during heating to produce carbon dioxide. The production of doughs by yeast tends to result in more flavor because it is a fermentation process.

Purpose of baking

Baking powder is a leavening (rising) agent used in most of the baked goods. It is used to make the cake light, fluffy and make the aerated crumb texture. If baking powder is not used, the end product will be dense.

Difference between cooking and baking refers primarily to the cooking of flour-based foods in which the heat of an oven sets their structures. Thus breads, cakes, muffins and loaves are all cooked by exposing them to particular temperatures that firm each specific dough in the center, with just the right degree of browning on the outside.

Advantages

- Cooking is meditative.
- A wide range of savory and sweet foods can be produced.
- Bakery products yield appetizing goods with eye-appeal and mouth-watering aromas. Thus it stimulates senses.
- Bulk cooking can be achieved with uniformity of color and degree of cooking.
- Baking ovens have effective manual or automatic controls.
- There is a straight forward contact for loading and removal of items.
- Nourishing activities feel good.
- Make other people happy.

Disadvantages

- Requires regular attention.
- Ovens are expensive to heat.

Aims and objectives of Bakery

1. Establish and maintain high standards of sanitation.

2. Exhibit a strong foundation of baking methodology.

3. Exhibit nutritional awareness and implement food-for-life principles.

4. Plan production of product and purchase, cost, and price product for profit.

5. Exhibit a solid foundation of techniques for food preparation, presentation and service, including competence in baking and pastry production, line work, and basic garde manger.

6. Use problem solving techniques in maintaining kitchen morale and building a team spirit.

7. Communicate clearly, both verbally and in writing.

8. Conform to professional standards in appearance, attitude, and performance.

9. Conform to established codes of ethics.

10. Demonstrate display techniques as they apply to hot and cold dessert presentations.

11. Demonstrate basic knowledge and skills for display pieces, including chocolate, marzipan, pulled sugar, and nougat.

12. Plan and present a grand pastry buffet.

13. Demonstrate working knowledge of the factors involved in setting up and operating a baking and pastry facility.

14. Demonstrate the ability to keep accurate food business records and understand the relationship between financial profits and good business ethics.

15. Demonstrate creativity and sound thinking in solving management problems in merchandising techniques.

16. Demonstrate a commitment to the profession through activities such as attending meetings, seminars, continuing education programs, and professional association memberships.

17. Develop skills in problem solving, decision making, and critical thinking.

Raw Materials in baking

In the process of baking, starch content in the food is processed usually decreased that provides the food a brown color which lends it an attractive and appetizing look. Some ingredients are required to prepare the bakery products. These ingredients are called "Raw Materials".

The ingredients are classified as:

- Essential ingredients
- Optional ingredients

Ingredients are classified according to their functions as:

- Structure builders
- Tenderizers
- Moisteners
- Flavorings
- Driers

Essential ingredients for bakery and confectionery products are:

- Flour
- Sugar
- Fat
- Eggs
- Yeast
- Salt
- Water

Optional ingredients are:

- Milk and milk products
- Dry fruits, Nuts and peels
- Flavors
- Chemicals
- Spices
- Chocolates
- Cocoa powder
- Corn flour
- Mixed fruit jam
- Custard powder
- Setting materials
- Colors.

Principles of baking

- Preheat oven to the required temperature.
- Weigh ingredients accurately.
- Understand ingredient function.
- Distribute foods evenly on greased baking trays to assist even cooking.
- Foods need to be placed in the appropriate position in oven.
- Even sized items on the same tray, small items, bake faster than larger items.
- Do not mix different items on the same tray.

Different ingredients have different purposes

Flour – Provides protein and starch

This forms the structure of baked goods.

Liquids – Aid flour to form the structure of baked product. Also, aids other ingredients in chemical process that occur.

Water, milk, fruits or vegetable juice, yogurt, and sour cream.

Eggs – Eggs make baked products tender, add flavor and richness and can help to bind mixtures together.

Flavorings – Chocolates, spices, herbs and extracts such as vanilla and almonds.

All types of food can be baked, but some require special care and protection from direct heat.

Events that occur during baking are:

- Fat melts
- Gases form and expand
- Microorganisms die
- Sugar dissolves
- Liquid evaporates
- Enzymes are inactivated
- Changes occur to nutrients
- Pectin breaks down
- Egg, milk and gluten proteins coagulate
- Starches gelatinized or solidify
- Caramelization occur

CHAPTER 2

EQUIPMENT NEEDED

OVENS

Baking ovens are major equipment for any bakery process. The major function of baking oven is to heat the wet dough, batter to a temperature where it becomes baked with desired texture and taste. Baking removes the moisture, which helps in improving the shelf life of

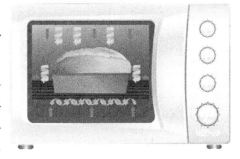

Fig. 1. Baking Oven

the baked products plus it kills any microbes in the dough at a higher temperature. Accessories to baking oven are circulating fan, steam extraction, chimneys, safety explosion doors, fire tube, burners, drive, temperature controller and indicators, fuel system with baking molds and wire bands. Bakery ovens selection also takes dimensions into consideration such as height, width, weight, chimney dimensions, foundation method, electrical wiring and automation.

Types of oven used in bakeries

Baking Ovens are of various types used for baking breads, cookies, puffs, biscuits, cakes, pizza, cream rolls and other bakery products.

Types of oven for baking can be classified by physical characteristics:

- Gas Oven
- Electric Oven
- Conventional vs Convectional Oven
- Steam Oven
- Self-cleaning Oven
- Warming drawers
- Rack Oven / Deck Oven
- Rotary Rack oven
- Travelling Oven
- Tunnel Ovens
- Swing Tray Ovens

Fig. 2 . Two tunnel oven

Con Tray Ovens

- Manual Loading
- Automatic Loading

Heating

- Direct Heating oven
- Indirect Heating Ovens
- Hybrid Ovens

Heat Transfer

- Conduction - movement of heat from one item to another through direct contact.
  - Pan placed over the burner
- Convection - transfer of heat through a fluid, which may be liquid or gas.
  - Hot air circulating in the oven

- Radiation - transfer of heat through waves that move from the heat source to the food.
  - Infrared cooking
  - Microwave ovens

Conduction (orange waves), Convection (yellow waves) and Radiation (red arrows)

Baking and Cooking Methods

- Dry-heat cooking uses air or fat and is the principal method to cook:
  - Batter
  - Dough
- Moist heat uses water or steam for cooking:
  - Fruits
  - Tenderizing Foods
  - Reducing Liquids

Different Types of Ovens

Ovens are a fundamental kitchen. It is a true investment, and it is important to know what baker's needs are and be able to articulate that into the type of oven, the baker is looking for. A baker need to know whether a baker want electric or gas, convection or conventional, and any add-ons a baker's desire.

Gas ovens

Similar to Cooktops, ovens fall under two main categories of energy sources – gas or electric. Gas ovens tend to be more expensive than electrical ovens of similar quality levels.

A common complaint about gas-powered ovens is that they tend to have hotspots and uneven heating throughout the oven. If a baker like baking, or have a problem with food browning, it is important to look into whether a baker would like a conventional or convection oven.

Electric ovens

Electric ovens work using heating elements placed on the inside walls of the oven. Electric ovens are also the easiest to use, easiest to clean and the easiest to achieve even cooking. They are also available in both convection and conventional varieties.

Conventional vs Convection ovens

Conventional ovens, also called traditional ovens, have no fans, and the air around the food is heated to cook it. Convectional ovens use fans to circulate that air, which usually cooks food faster and more evenly.

Steam ovens

Steam ovens are a less versatile, but more nutritious way to cook food. Using steam to cook food, means that less nutrients are lost and there is no need to use oil or butter. However, it is not great for getting any variation of color or texture.

Self-cleaning ovens

Self-cleaning ovens or Pyrolytic ovens are a luxurious way to cut down on a tedious maintenance job, but also provide comfort. Whilst cleaning, Pyrolytic ovens lock themselves until they reach a very hot 500°C, which turns any food remains of ash, which just sweep away when it is done. Cremating forgotten food remnants is possibly the most satisfying ways to clean an oven.

Warming drawers

This drawer works similarly to a teapot and is perfect for heating up the plates, keeping side dishes from going cold, or preserving the perfect warmth that is freshly baked bakery products.

Principles of baking oven

Heat and temperature are not the same and should not be confused. It is relatively easy to measure temperatures in an oven, but much more difficult to measure heat, or heat flux, which is the rate at which heat is being transferred. Heat is transferred much more effective if the air is moving near the dough piece at a given temperature.

Nearly all bakery products are now baked in a band or travelling ovens with several independently controlled zones. This means that oven conditions such as temperature, movement and humidity of the atmosphere may be altered during the course of the baking period. Baking times for bakery products are quite short, ranging from 2.5-15 minutes. It is not normally possible to change, quickly, the temperature of a static or reel oven so the results of baking in these ovens compared with that in travelling ovens are often very different.

The conditions needed for different types of bakery products are not the same because the way in which the structure is developed and the amount of moisture that must be removed depends on the richness (level of fat and sugar) of the recipe. The baking requirements for different types of bakery products will be considered later.

There are four major changes to the dough piece which can be seen as it is baked,

1. A large reduction in product density (the dough gets thicker) associated with the development of an open, porous or flaky structure.

2. A change of shape associated with shrinkage or spread and increase of thickness.

3. A reduction of moisture level, to between 1-4%.

4. A change in surface coloration (reflectance).

Although these changes are thought of as being distinct and sequential, broadly in the above order, as the product passes through the oven, it will be shown that there is considerable overlap and coincidence of these physico-chemical changes.

DOUGH MIXER

Working principle and method

- Put the dough smooth corners of workplace, plugged in, according to switch to start the motor, working free first, pay attention to the direction of rotation of the mixing with machine, gear same direction of the arrow on the cover, check whether the transmission system is normal and a reliable grounding machine shell.

- To the surface, stirring with besmear brushes a small amount of cooking oil, as to avoid the dough adhesive.

- First combine the flour, with the proportion of water 100:40-50, if the surface is relatively hard, midway can slowly add waste, if the dough with agitator rotational, along the surface of a pipe side scatter.

- Generally, with 3 to 8 minutes, the surface can be reconciled, turn the power off and pull out the dowel pin, will face dough, turn 90 degrees, with positioning pins, again started with the switch, turn back mixing with dough can be automatically removed. After the 5, and, in a timely manner to do a good job of cleaning maintenance.

It is generally made to order, but available capacities vary from 25 kg to 35 kg. Most commercial modules are heavy and should be fitted with sturdy rollers for easy movement. The stainless steel bowl and beater should be washed and cleaned after every use and the machine should be serviced regularly.

EGG BEATER

- A hand tool that is used to manually mix and beat eggs or other similar ingredients, such as sauces, batter, egg whites and dressings.

- This kitchen stainless steel with stainless blades all of which can be easily cleaned.

- An egg beater is most often used for mixing and blending foods quickly, easily and without much preparation or cleanup.

In text questions

1. Self-cleaning ovens is otherwise called as _____ ovens.

2. _____ ovens are a less versatile, but more nutritious way to cook food.

3. Conventional ovens, also called _____ ovens, have no fans, and the air around the food is heated to cook it.

4. Electric ovens work using _____ placed on the inside walls of the oven.

5. _____ ovens use fans to circulate that air, which usually cooks food faster and more evenly.

6. Define Baking.

7. Explain the historical background of baking.

8. What are the qualities a good baker should have?

9. What is the purpose of baking?

10. Write five objectives of baking.

11. What are the basic principles of baking?

12. What is the difference between cooking and baking?

13. Why is baking so important?

14. What happens during baking?

15. Classify the raw materials and discuss the functions of the ingredients used in bakery.

16. What are the fundamentals of baking?

17. Outline the demerits of baking

18. List the raw materials used to prepare bakery products.

19. Classify ingredients according to the functions of baking.

20. Categorise baking ovens.

21. Write the working principle of baking ovens.

22. Mark the working principles of dough mixer.

23. Brief on the working principles of egg beater.

CHAPTER 3

WHEAT

Flour obtained from wheat plays a vital in the manufacture of bakery products. Wheat is the most important cereal among all grains. The quality of wheat is determined by several factors depending upon the following conditions.

1. Soil

2. Quality of seeds

3. Climate

4. Manure

5. Farming techniques

The physical structure of wheat consists approximately of:

• Bran

• Outer skin or epidermis

• Second skin or epicarp

• Third skin or endocarp 15%

• Fourth skin or testa

• Fifth skin or Aleurone layer

• Germ 2.5%

• Endosperm 82.5%

In the production of flour for baking both bran and germ are removed during the milling process. Removal of bran is essential because the sharp edges of bran will tend to cut the cell structure of the loaf during proofing thereby affecting the volume of bread. Fermentation and proofing are the processes which take place after mixing when the dough is kept under controlled conditions. Bran is high in nutritive value, and is mostly used for animal feed. The germ is removed from the wheat during milling. Because the germ is having high oil content which will affect the keeping quality of flour.

From 100 kg of wheat, 72% extraction is known as 100 percent straight flour comprising of all the streams. The word 'stream' is used in milling technology, which means flow of flour in a continuous succession of different stages in the flour milling process.

The remaining 28 percent consists of bran and shorts, which is akin to resultant Atta. This is mostly used for feed and industrial purposes.

During milling, the endosperm particles must themselves be categorized according to their size. The manner in which these particles are separated is called "separation". Extraction refers to the percentage of flour which has been extracted from wheat kernel.

Therefore, an average flour, depending upon extraction and separation, will consist of the following:

Starch	:	70%
Moisture	:	14%
Protein	:	11.5%
Mineral (ash)	:	0.4%
Sugar	:	1%
Fat (liquid)	:	1%
Others	:	2.1%

Wheat is classified in various methods such as

1. Type

2. Color

3. Hardness

According to the type they are classified as

1. Triticum Aestivum (also called hard wheat)

2. Triticum Compectum (also called soft wheat)

3. Triticum Durum (also called durum wheat)

The Triticum Aestivum wheat flour contains more proteins. This flour is used for the production of bread.

The Triticum Compectum wheat flour contains low protein. So this flour is used for the production of biscuits, cakes and pastries.

The Triticum Durum wheat is mainly used to prepare and macaroni which is used chiefly in making elementary pastes such as semolina, macaroni, spaghetti, noodles, etc and red durum which has very little value for milling and is used principally as feed. All flours are not of the similar composition percentage-wise. Several factors such as the effects of climate, breed of seed, the type of seed, the type of wheat blended and proportions of the wheat used during blending and the storage period of wheat will also affect the quality of flour.

According to the Color, they are

1. Red wheat

2. White wheat

This color variation is due to the environmental factors.

According to the Hardness, it is classified into

1. Hard wheat

2. Soft wheat

Examples of Hard wheat are

1. Hard red winter
2. Hard red spring
3. Durum

Examples of Soft wheat are

1. Soft red winter
2. Soft red spring.

Hard wheat

Bakery products are made from this type of wheat flour because hard wheat flour contains the following characteristics:

1. More protein.
2. More Water Absorption Power (WAP)
3. Good mixing capacity that is easy to mix
4. Good Fermentation tolerance
5. Good gas retention power

   Hence it is mainly used for yeast products (Eg. Bread).

Soft wheat

Soft wheat flour contains the following characteristics

1. Low protein
2. Low Water Absorption Power (WAP)
3. Poor mixing capacity
4. Poor fermentation tolerance

   Hence it is mainly used to make biscuits, cakes, and pastries.

Identification of flour based on its characteristics

Different types of flour are used for different types of end products. Flours are identified as First Patent, Short Patent, Medium Patent and Long Patent. Characteristics of these flours are determined by percentage of separation obtained from 72% extraction.

- First Patent constitutes 70% separation from 72% extraction. First Patent is used as cake flour and is obtained from soft wheat.

- Short Patent constitutes 80%. Short Patent is used for premium brands of breads

- Medium Patent 90% Medium Patent is used for featured brands of breads.

- Long Patent 95% separation from 72% extraction. Long Patent is used for competitive brands of breads.

Cake flours should contain less than 10% protein and 0.4% ash, and should have low absorption. Ash content of flour is considered as a measure of the amount of separation of the flour from a particular wheat blend, but is not a reliable index of baking industry.

Quality of proteins is a more important factor in determining baking properties of a flour than the protein quantity. Loaf potentialities are determined by the gluten quality and quantity.

Structure of wheat

The kernel of wheat is a storehouse of nutrients. The wheat is classified into 3 principle parts. They are :

1. Bran

2. Germ

3. Endosperm

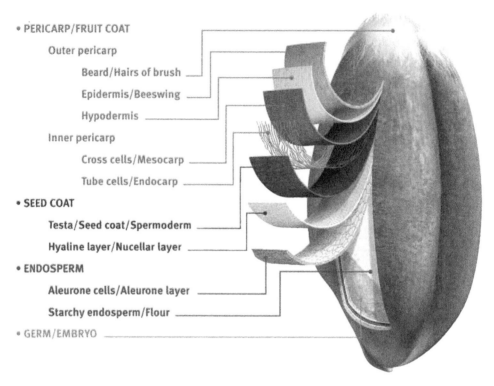

- PERICARP/FRUIT COAT
    Outer pericarp
        Beard/Hairs of brush
        Epidermis/Beeswing
        Hypodermis
    Inner pericarp
        Cross cells/Mesocarp
        Tube cells/Endocarp
- SEED COAT
    Testa/Seed coat/Spermoderm
    Hyaline layer/Nucellar layer
- ENDOSPERM
    Aleurone cells/Aleurone layer
    Starchy endosperm/Flour
- GERM/EMBRYO

Fig. 3. Cross section of wheat

Bran

Wheat contains 15% of the bran. It is the outer portion of the wheat. It has several layers. They are

1. Epidermis
2. Epicarp
3. Endocarp
4. Testa
5. Aleurone layers
6. Skin or Aleurone cells

The color of wheat is due to the testa. This layer protects the endosperm. The Aleurone layer has small cells. These cells have enzymes, which converts starch into sugar and it gives softness to the flour.

The bran contains more nutritional value, even though it is removed during the milling process, as the sharp edges may cut down the gluten. During milling, bran is removed and is used as animal feed.

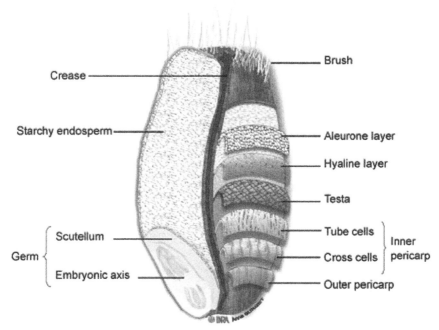

Fig. 4. Structure of wheat

Germ

Wheat contains 2.5% of germ. It is the sprouting section of the seed. During milling, the germ is removed because it has a fat content that will spoil the flour quickly.

Endosperm

Wheat contains 82.5% of endosperm. Although primarily, Starch it contains the following nutrients.

1. Protein

2. Pantothenic acid

3. Riboflavin

4. Niacin

5. Thiamine

During milling, endosperm is separated from the bran and germ.

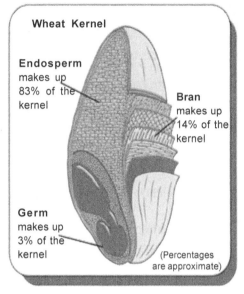

Fig. 5. Composition of wheat kernel

In Text Questions

1. The _____ wheat flour contains more proteins and it is used for the production of bread.

2. The kernel of wheat is a storehouse of _____.

3. Wheat contains _____ % of endosperm.

4. During milling, _____ is separated from the bran and germ.

5. Classify wheat.

6. List the type of wheat flours and bring out the characteristics of good quality flour.

7. Explain the physical structure of wheat with a neat diagram.

8. How the quality of wheat is determined?

9. Explain wheat milling process in detail.

10. How wheat is classified?

11. How are the characteristics of flours determined by percentage of separation obtained from percentage of extraction?

12. Enumerate the Constituents of flour.

13. Brief on Hard wheat.

14. Write a note on Soft wheat.

CHAPTER 4

FLOUR

The English word "flour" is originally a variant of the word "flower" and both words derive from the Old French *fleur* or *flour*, which had the literal meaning "blossom", and a figurative meaning "the finest".

Flour is a substance, generally, a powder made by grinding raw grains or roots and used to prepare many different foods. Cereal flour is the main ingredient of bread, which is a staple food for many cultures. Wheat is the most common base for flour.

Cereal flour consists either of the endosperm, germ and bran together (whole-grain flour) or of the endosperm alone (refined flour).

Different kinds of flour in India

Flour provides the structure in baked goods. Flour contains a high proportion of starches, which are a subset of complex carbohydrates also known as Polysaccharides. "Bleached flour" is any refined flour with a whitening agent added. "Refined flour" has had the germ and bran removed and is typically referred to as "white flour".

Wheat flour

Wheat flour contains proteins that interact with each other when mixed with water, developing gluten. It is this elastic gluten framework which stretches to contain the expanding leavening gases during rising. The

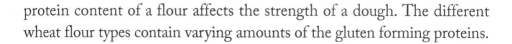

protein content of a flour affects the strength of a dough. The different wheat flour types contain varying amounts of the gluten forming proteins.

Bread flour

Bread flour is a hard wheat flour with about 12 percent protein. Bread flour is used for yeast raised bread because the dough yields more gluten than dough made with other flours.

Cake flour

Cake flour is a soft wheat flour that is 7.5 percent protein. The lower gluten content causes products to have a tender, crumblier texture that is desirable in cake.

All-purpose flour

All-purpose flour is blended during milling to attain a protein content of 10.5 percent. This medium protein flour can be utilized for all baking purposes. All-purpose flour is formulated to have a medium gluten content of around 12 per cent.

Whole wheat flour

Whole wheat flour may be substituted for part of the white flour in yeast and quick bread recipes, but the volume of the finished product will be decreased. Whole wheat flour contains the nutritious germ and bran as well as the endosperm of the wheat kernel. Bran particles cut through the gluten during mixing and kneading of bread dough, resulting in a smaller, heavier loaf.

Wheat germ

Wheat germ, though not a flour, is often used in place of part of the flour in recipes for flavor and fiber. Protein, vitamins, minerals, and polyunsaturated fats are concentrated in the germ of grain kernels. The following non-wheat grain products are often used in baked goods. They are rich in protein but most do not have the potential for developing gluten.

Rye flour

Rye flour is often used in combination with wheat flour for bread.

Triticale flour

Triticale flour is a hybrid of wheat and rye. It has an average protein content higher than that of wheat flour. In yeast bread dough, triticale flour has better handling properties than rye flour because it will form gluten, but does not handle as well as wheat dough. For a good quality dough, ferment yeast dough made with triticale flour for a shorter period than wheat flour dough.

Oat flour

Oat flour has a relatively high protein content, 17 percent, but does not form gluten. Oat flour can be substituted for as much as 1/3 of wheat flour in bread.

Corn meal

Corn meal is coarsely ground dried corn. Corn flour is more finely ground corn. Both corn flour and corn meal contain 7-8 percent protein on a dry basis. Neither corn meal nor corn flour will form gluten. A grainy texture in cornbread can be avoided by mixing the cornmeal with the liquid from the recipe, bringing to a boil, and cooling before mixing with the other ingredients.

Rice flour

Rice flour has about 6.5-7 protein and does not form gluten. For people who do not tolerate gluten, rice flour is an acceptable substitute for wheat, barley, rye or oat flours.

Potato starch flour

Potato starch flour, other non-gluten forming flour is usually used in combination with other flours. It has a mild potato taste.

Soy flour

Soy flour contains 50 percent protein and is used primarily to boost the protein content of baked goods. This flour cannot form gluten and does not contain starch. Its use in large amounts affects the taste of baked goods and causes them to brown quickly.

Constituents of flour

1. Starch
2. Protein
3. Moisture
4. Ash
5. Sugar
6. Fat or Lipid
7. Others (Enzymes – Alpha and Beta amylase)

Starch

- Starch is not soluble in water until the starch is heated to about 140°F with 6 times of its weight of water.

- Water Absorption Power (WAP) of the flour mainly depends upon the damaged starch. Enzymes (alpha and beta amylase) act only on damaged starch to produce sugar for the yeast during fermentation. The damaged starch should not be more than 7% to 9% for bread making. Damaged starch is not essential for cake or biscuit making.

- Hot bread directly from the oven cannot be sliced immediately because the starch is not sufficiently stable and must be allowed to retrograde (slightly harder). When the bread cools down starch cells shrink and becomes rigid so that the bread can be sliced easily.

Moisture

- An ideal moisture content of flour is 14 %. The source of moisture is from the tempering or from the package materials or from the humidity. If more moisture is in the flour, it will reduce the storage life and will induce insect infestation and it may get fungus and bacteria and also it will reduce the WAP of the flour. This will result in less yield during production.

Protein

Flour contains soluble and insoluble proteins. Flour protein consists of

1. Albumin
2. Globulin
3. Gliadin
4. Glutenin

Soluble proteins are useful in providing nourishment to yeast during the fermentation process for its growth and reproduction. The insoluble protein and glutenin form a rubbery material when water is added to flour, so when it is mixed and kneaded well, a rubbery material is developed. This is called gluten. The quality of flour is determined by gluten content. If gluten content is more in flour, then it is suitable for high structured products like bread. This bread making flour should have gluten from 10-11.5 %.

Ash

The source of ash content in flour is from the bran. If the flour contains more ash, it means it has more bran. Too much ash gives a dark color to the flour and also cuts the gluten. Flour with higher ash content will not retain as much as a gas during different stages of processing and this affects the volume and gives poor texture to the product.

Sugar

Naturally flour contains a small quantity of sugar. That is sucrose and maltose. It is used as yeast food to produce CO_2.

Fat or lipid

This should not be more than 1% in flour. It contains coloring pigment carotene, which gives color to the flour. There is a higher quantity of oil/ fat in the low grade flour than in high grades. The fat or oil when separated from the flour is a pale yellowish liquid without taste or smell.

Enzymes

Flour contains diastatic enzymes. They are alpha amylase and beta amylase. These enzymes hydrolyze starch and convert it into simple sugar. During fermentation, the simple sugar is used by the yeast to produce alcohol and CO_2. The gas production depends upon the amount of enzymes found in the flour. Indian flours have less alpha amylase. These enzymes are necessary for producing good quality bread. In rain damaged wheat, the enzymes will be in excess. If the bread is made out of this flour, the bread will have dark crust color and sticky crumbs. If these enzymes are less, the bread will have poor volume and dull crust color.

Classification of flour

Bakers use two types of flour. They are

1. Hard flour
2. Soft flour

Hard flour contains 11.2-11.8% protein, 0.45-0.50% ash, 1.2% fat and 74-75% starch. Soft flour contains 8.4-8.8% protein, 0.44-0.48% ash, 1% fat, and 76-77 % starch.

Types of flour in bakery

1 High ratio flour
2 Whole wheat flour
3 Whole meal flour
4 Self-rising flour

High Ratio Flour

This flour is also known as special cake flour. This absorbs high liquids, fats and sugars than normal flours. It is normally manufactured for special order and it is used in special recipes. This flour is normally blanched with chlorine gas.

Whole Wheat Flour

It is milled for whole wheat grain and no bran or germ is removed during milling. When using this flour, it requires more liquid than mentioned in the recipes.

Whole Meal Flour (Brown Flour)

In this mixture of refined flour, the content of bran and wheat germ are more. It can also be made by combining white and whole wheat flour.

Self-Rising Flour

It contains a certain quantity of baking powder. If we use this flour, we should reduce the baking powder quantity from the given formula.

Characteristic of good quality flour

1. Color
2. Strength
3. Tolerance
4. High absorption
5. Uniformity

Color

Flour should be creamish white in color. A good quality flour will reflect the tight when it is seen under the light. Bleaching the flour helps to get the color.

Strength

There are two types of flour.

1. Strong
2. Weak

It depends upon gluten quantity and quality present in the flour. For making bread, strong flour is preferred and weak flour is preferred for making cakes and confectionery products.

Tolerance

This is the ability of the flour to withstand the fermentation and/or mixing process in excess of what is normally required to mature its gluten properly.

High Absorption Power

This means the ability of the flour to hold the maximum amount of water. If the flour has less WAP, the bread will not be of good quality and will have less yield.

Uniformity

If the flour used is not uniform, then the quality of the product will differ. So constant monitoring and adjustments are required to get a satisfactory result.

Functions of flour in bakery products

- It acts as a binding agent and an absorbing agent.
- It is important for flavor.
- It adds nutritional value.
- It builds the structure.
- It is the backbone of the baked products.
- It affects the shelf life and quality of products.

In Text Questions

1. _____ flour is often used in combination with wheat flour for bread.

2. Triticale flour is a hybrid of _____ and _____.

3. In _____ dough, triticale flour has better handling properties than rye flour because it will form gluten.

4. _____ flour has about 6.5-7 protein and does not form gluten.

5. _____ of the flour mainly depends upon the damaged starch.

6. _____ and _____ act only on damaged starch to produce sugar for the yeast during fermentation.

7. Flour contains _____ enzymes.

8. Classify the types of flour in bakery.

9. Differentiate hard flour and soft flour.

10. Describe the role of flour in baking.

11. Enumerate the factors affect the quality of flour.

12. What are the characteristics of good quality flour?

13. Enumerate the functions of flour in bakery products.

CHAPTER 5

FATS AND OILS

INTRODUCTION

A recent survey found 20 different varieties ranging from traditional butter with 81.5% fat to a low fat spread containing only 40% fat. If a fat is to be classed as a margarine under FSANZ (Food Standards for Australia and New Zealand) it must consist of at least 80% fat. If it has less than this amount it is classed as a spread. Most spreads are 70% or 75% fat. Table 1. shows typical fat contents of butter, margarine and spreads. It can be noted that the amount of salt varies considerably.

Table 1. Comparison between Butter, Margarine (Polyunsaturated), Spreads and "Light" Products

Types of Fat	% Fat	% Moisture	% Salt
Butter	81.5	16	2.4 – 2.6
Polyunsaturated Margarine	80	16	1.0 – 2.0
Typical Spread	70 – 75	20 – 25	1.0 – 1.8
Light Spread	60	35 –40	1.0
Low Fat Spread	40	60	1.0

All butter, margarine and spreads are water in oil emulsions. Margarine and spreads contain added emulsifiers such as lecithin and monoglycerides to aid in the emulsion preparation. Butter in contrast, contains milk fat lecithin, a natural emulsifier. Spreads are originally designed for table use and not specifically for baking. Fat plays a different role in each baking application.

Types/Variations of fat used in baking

Shortening: Made from 100% vegetable fat, it is solid at room temperature. Acquired by the hydrogenation of vegetable oil, shortening provides pastries their flakiness, and gives cakes or cookies a lighter feel if creamed with sugar to trap air. Shortening affords breads with stability, preventing airflow within the loaf during baking, and gives a desirable smooth mouth feel and flavor.

Butter: Consists of 80% fat and is made from cream. The remaining 20% is water combined with milk solids. Butter imparts flavor and a greasy mouthfeel to all baked goods due to solubility at body temperature.

Clarified Butter: Butter that has been heated to eliminate the sediment of milk solids therefore turning a clear color.

Margarine: Consists of 80% vegetable fat, can be used interchangeably with butter.

Oil: The original version of vegetable oil from soybean, canola or corn sources. It is used in some muffin, bread and cake recipes. Pastries are utilizing oil rather than the solid fat result in a mealy texture rather than flaky. In regard to cake recipes, interchanging oil for solid fat results in a heavier texture unless counteracted via increasing sugar or egg. To replace oil for butter or margarine, use 7/8-cup oil for one cup of butter or margarine.

Ghee: Class of clarified butter with a subtle yellow color and rich nutty flavor, used as a substitute for pure butter in many cultures. Vegetable ghee, made from various vegetable oils, is more commonly used than ghee made via butter.

Palm Oil: The main source of trans fat-free shortening; solid in nature at room temperature and made from palm oil.

Cocoa Butter: Pale yellow, pure, edible vegetable fat extracted from cocoa beans. Used in chocolate chips.

Olive oil: Fat obtained from grinding whole olives and extracting the juice. Used in focaccia bread and also in any baked good recipe. Can be substituted for vegetable oil.

Lard: Pig fat has a very high smoke point, making lard ideal for culinary usage. In baking, lard is used in cookie production, pie crusts, and cakes.

Role of fat in baking applications

- Shortening reduces the toughness of dough.
- It improves dough for machining and sheeting by lubricating the gluten.
- Controls the flow of dough
- Gives shorter bite to the goods
- Enhances the product flavor and taste
- Fat provides nutrition. Fat is a good source of energy due to the high level of calories.
- Provides extensibility in bread dough.
- A small amount of fat in a yeast dough supports the gluten to stretch, producing a loaf with greater volume.
- Improves texture and grain in bread.
- Helps to retain the gases released during baking thus ensuring a well risen loaf which will have a soft crumb and will stay fresh longer.
- Gives softness to the product.

- Helps in aeration-The fat holds the air that is trapped in during the creaming stage and when egg is added they assist the fat in holding the air.

- Aids in laminating the product- Usually a "toughened" plastic type fat that can withstand the rolling and folding process of lamination. It allows the layers of fat and dough to be built up.

- Helps to increase the keeping qualities - Fats and oils help to extend the shelf life of a product.

In Text Questions

1. FSANZ is ——————-.

2. ————- that has been heated to eliminate the sediment of milk solids so as to turn into a clear color.

3. Margarine and spreads contain ———- and ——— to aid in the emulsion preparation.

4. —————— aids in laminating the products.

5. Butter is obtained from _____ by the churning method.

6. Can use _____ cup oil for one cup of butter/margarine.

7. _____ converts the oil into shortening.

8. Butter contains _____ % fat in it.

9. Lard is obtained from _____.

10. Classify different types of fats used in bakery products.

11. Enumerate the functions of fat in baking

CHAPTER 6

EGG

Egg is the most chief raw material used in bakery products. Besides the egg of ducks, geese and turkeys can be used for production. As this egg has different characteristics, bakers prefer hen eggs only.

Composition of Egg

Egg is composed of the following three parts.

Shell 12%

White albumen 58 %

Yolk 30%

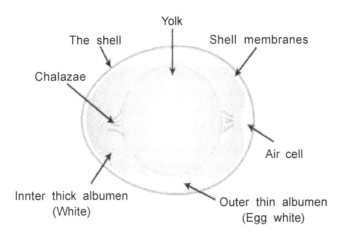

Fig. 6. Structure of Egg

There are different grades and sizes of egg available in the market. The average egg size starts from 45 g to 70 g.

Types of Egg

There are two types of eggs used in bakeries.

- Shelled eggs
- Frozen eggs

Shell eggs, both fresh and cold storage, are sometimes used in bakery products. Dried eggs are also used and their usage is increasing. By far the largest amount is in the form of frozen eggs. Frozen eggs are sold as whole eggs, whites and yolks. In frozen whole eggs, the proportion of yolk and white may differ from the shell eggs - for often frozen whole egg is "fortified" by the addition of yolk. As far as shelled eggs are concerned, quality eggs must be used to ensure quality of the finished products.

Advantages of Frozen Eggs

- They save time and labor because breaking and separating are not necessary.
- They avoid waste through spoilage and other causes.
- An adequate supply of uniform quality is assured. Uniform quality means better results.
- Frozen eggs are sold as whole egg yolks, whites or specials which contain whole eggs and extra yolks.

Quality Test

The egg quality can be judged by different methods.

Candling method

The quality of whole egg is generally evaluated by the candling method. Here, the egg is held before a bright light in a dark room and its contents observed through the shell. If the center portion is dark and the rest is opaque, then it will be a good egg. If the rest of the egg parts are semitransparent, it indicates the beginning of deterioration and if it is formed nontransparent, the egg will be already spoilt.

Moreover, the egg is given a quick twirl and the extent of the resultant motion of the yolk is observed. If the movement of yolk is restricted, it will be a good egg because fresh egg contains thicker white compared to stale egg.

Salt Solution Test

Keep the egg in a vessel filled with 10% salt solution. If the egg sinks, its quality will be good. If it floats on top, it will be spoiled because the gas produced due to protein decomposition makes it lighter. However, on the basis of the experience, one can decide based on the speed of sinking.

Eggs may be further classified into:

Composition

1. Whole Egg
2. Egg Yolk
3. Egg White

Whole Egg: In calculating the amount of eggs to be used in a recipe or formula, one can assume that the whole egg is approximately 75 percent moisture, the remainder being solids.

Yolk: The yolk of an egg contains most of the fatty material in a finely emulsified state. The approximate amount of lecithin fat in the yolk is 7 - 10% of total fat content. Yolks are used for improved creaming, greater volume etc. Although the yolk appears to be almost semi-solid, it contains almost 50% water.

Egg White: Egg white contains approximately 86% moisture. The whites are either firm or fluid in nature. The whites close to the yolk are generally firm, while the portion closer to the shell is fluid.

Fresh Egg: It refers to eggs that have recently broken, or separated from the shell and placed in cans, these are usually preserved by freezing.

Frozen egg: The eggs are quick frozen at -10 °F to -15 °F and may be stored for a long period at 0°F or below without spoiling.

Dried Egg: Eggs are dried by spraying into a heated chamber (160 - 170°F), the moisture is almost completely removed.

Soya Flour: Can replace eggs - can use 30% based on flour.

The composition of eggs is shown in the following:

Table 2. Composition of eggs

S.No.	Components	% Whole Egg	% Egg Yolk	% Egg White
1.	Moisture	73.0	50.0	86.0
2.	Protein	14.0	17.0	12.0
3.	Fat	12.0	31.0	0.2
4.	Sugar	0.0	0.2	0.4
5.	Ash	1.0	1.5	1.0

Functions of Eggs in Bakery Products

1. **Increase Nutritive Value:** Eggs are high in nutritional value and their use in baked goods improves the value of these products as food. Eggs are an important source of the necessary minerals iron, calcium and phosphorous. While milk is rich in calcium and phosphorous, it is low in iron. Iron exists in very small quantities in most foods, but egg yolk contains a relatively large amount in a

form which the human body can assimilate readily. Egg protein is a complete protein, capable of supplying all of the essential amino acids required to maintain growth and good health. Both the protein and the fat, which is in the yolk, are of a nature to be readily assimilated by the body. In addition, eggs supply important amounts of vitamins A and D, thiamine and riboflavin.

2. **Improve Flavor, Texture and Eating Quality:** Eggs have an odor which some people consider desirable in the baked products.

3. **Aid in Producing an Appetizing Color in both Crumb and Crust:** The yolk of the egg provides the desirable yellow color which gives the cake a rich appearance.

4. **Acts as a Binding Agent to Hold the Various Ingredients Together:** Example - custards

5. **Contributes Emulsifying Action:** Example - lecithin in yolk

6. **Produces a Shorter Crumb:** Because of the fat and other solids of the eggs, the product has additional fat and tastes sweeter. Eggs also provide shortness in the mix, enabling the mix to be handled easily.

7. **Improves Keeping Quality:** Because egg contains 75% moisture and natural ability to bind and retain moisture, they retard staling. This is especially true of products made with additional yolks.

8. **Aids in Leavening, Especially in Products Such as Angel Food:** The foam from whipped or beaten eggs entraps air bubbles which expand when heat is applied. In the mix, they improve creaming, increase the number of air cells formed and coat these cells with a fat which permits further expansion of the air cells. In baking, the air cells expand further and the partial evaporation of moisture in the form of steam, increases leavening. When whipped, as for sponge cakes, the foam formed by the eggs affects the leavening.

USE OF EGGS IN BAKERY PRODUCTS

- Egg acts as a Leavening Agent

- When the egg white is beaten, the foam consists of many small air bubbles, each surrounded by a film of egg protein. The ability of egg white to foam is due to its low surface tension which results in a concentration of solids at the surface.

- The mechanical action of beating and contact with the thin protein film with air, partly coagulate the protein and make the foam stable.

- On baking, the air bubbles expand with heat and the protein film is sufficiently elastic to stretch.

- As the batter or meringue reaches higher temperature, the protein coagulates entirely, losses its elasticity and sets to a firm structure.

- The volume and stability of egg white foams are influenced by a number of factors.

- The more viscous or thick the egg white is the longer it takes to form a foam, and the lower will be the volume.

- On the other hand, the stability of the foam is somewhat greater. This effect may not be due to the egg so much as to the fact that thick egg white may not be carried along by the beater instead of mixing readily.

- The demand of thick egg whites is probably due to the fact that in the case of shell eggs, the ability of the white of a freshly cracked egg to stand high when it is poured on a plate is an indication of freshness.

- In old eggs the white has thinned out, and flows more readily.

- Obviously, in frozen or dried eggs, which are thoroughly mixed or beaten before processing, this test has no meaning.

- Other factors which influence the volume and stability of the foam produced by egg white are the presence of salt, sugar and acid and the conditions under which the white is beaten.

- Salt and sugar, within certain limits of concentration, stability by bringing the protein to its isoelectric point where it coagulates more rapidly.

- This is the reason for using such acid ingredients as cream of tartar, calcium acid phosphate or lemon juice in a meringue.

- Salt and acid stabilizers are most effective when added at the foaming stage, early in the process.

- Sugar has its greatest effect also, when added at this stage, but lengthens the time required for beating.

- Egg white of good quality should be entirely free from fat of any kind. The presence of fat results in a form of low volume, fine structure and low stability.

- Since surface tension is lowered to raise in temperature, egg white whips more rapidly at room temperature than when cold.

- The type of beater affects the foam also. If it is too coarse, the foam will have large bubbles and will be less stable.

- When egg yolk is whipped both the kind of protein it consists of and presence of the large amount of natural fat, which it contains give the foam low volume, a fine structure, and less stability than egg white when cold.

- However, egg yolk has good emulsifying properties which aid in incorporating the other ingredients, and in keeping fat particles dispersed in a batter. As stated above, the properties are due to the presence of lecithin, a natural constituent of yolk.

- On baking, yolk coagulates to more stable and firmer structure than the white, and forms a structure with thinner cell walls.

In text questions

Fill in the blanks:

1. Eggs have _____ nutritive value.

2. Yolk of egg contains _____ %of fat content.

3. Eggs can be replaced by _____ flour.

4. Whole egg has _____ %moisture in it.

5. In custard, eggs act as _____ agent.

6. Discuss the different types of eggs in bakeries.

7. How egg acts as a leavening agent?

8. Why frozen eggs have become popular? How it is made?

9. How would baker check the freshness of an egg?

10. Give the composition of Eggs.

11. State the functions of Eggs in bakery products.

12. Describe the use of eggs in bakery products.

CHAPTER 7

MILK

Milk is an important moistening agent for bakery and confectionery products. Fresh milk is a white fluid with a pleasant odor and a slightly sweet characteristic taste. Milk contains fat, carbohydrate (lactose), protein (casein), minerals, vitamin, etc. Milk is an emulsion in which milk fat is dispersed in water. Emulsions are colloids in which both disperesed phase and dispersed medium. So milk is an emulsion in which liquid is dispersed in water. The sugar and fat in milk help tenderize and moisten the baked good, while adding flavor.

Milk is available in many forms and is divided into three.

1. Liquid form

2. Concentrated milk

3. Dry form

Fresh milk, pasteurized milk and skimmed milk, all come under the liquid form. Fresh milk is a natural form as it comes from the cow. Fresh milk should be stored at 40°F or below. When milk has to be kept for long period, it is pasteurized and the bacteria are not completely killed but their multiplication is prevented. The milk that remains after removing the butter fat from the whole milk is termed as skimmed milk. Whereas, the butter fat present in milk keeps the product moist for longer time and improves shelf-life of the product.

Condensed milk is concentrated up to a syrupy thickness in vacuum pan set by heating at temperature well below protein coagulation at 70°c. Evaporated milk is the whole milk with a part of its water removed by evaporation up to desired standard level (example: less than condensed milk)

Milk powder is an example of dry form milk, it is made by heating milk below the protein coagulation temperature and thereby removing the moisture. Milk powder has extended shelf life and consistent qualities, when compared to fresh and other forms of milk.

Functions of Milk

- It gives the nutritional value of the products.
- It has a toughening effect on flour protein, which improves the gas retention power of dough.
- It helps in producing soft and silky structure.
- It contains lactose. This lactose helps to give crust color to the product and it improves the water retention power.
- It gives flavor and taste to the product.
- It improves the texture to the products.
- Milk has a buffering action on dough but excessive fermentation may produce undesirable quantities of lactic acid, which may break down the gluten resulting in bread, having a very coarse and dark crumb and unpleasant sour taste and flavor.
- It keeps the product tender.
- The butterfat present in milk keeps the product moist for a longer time and it improves the shelf-life and also imparts unique flavor to the products.

Advantages of using Milk solids in Bread Production

There are several advantages that could be resulting from adding milk solids in the bread dough. These are listed below:

1. Increased Absorption and Dough Strengthening

2. Increased Mixing Tolerance

3. Longer Fermentation

4. Better Crust Color

5. Better Grain and Texture

6. Increased Loaf Volume

7. Better keeping Quality

8. Better Nutrition

Functions of NFDM (Non-Fat Dry Milk)

1. Milk and milk products impart a high moisture absorption capacity to dough, causing an increase in dough handling during processing.

2. Increase buffering capacity during fermentation and thereby prevent rapid and excessive acidification.

3. Facilitate better control of amylase activity.

4. Improve tolerance to promote.

5. Facilitate moisture transfer during gelatin of starch.

6. Minimize the effects of over mixing.

7. Enhance flavor development and crust color formation.

8. Improve toasting characteristics.

9. Strengthen crumb structure and texture.

10. Improve moisture relation and retard staling process.

11. Enhance nutritional value.

12. The protein efficiency ratio (PER) of bakery goods is significantly improved when milk products are used in the formula. The reason for this nutritional improvement is that wheat flour protein is deficient in lysine and tryptophan, both being essential amino acids.

In text Questions

1. The_____ present in milk keeps the product moist for a longer time.

2. The sugar and fat in ——— help tenderize the baked good, while adding flavor.

3. Milk is an emulsion of _____and_____.

4. The milk protein is known as_____.

5. Milk improves_____color in bread.

6. The milk sugar is known as _____.

7. Milk has maximum mineral contents in terms of _____ and_____.

8. Milk should be stored at _____deg for below.

9. Explain the process of making the liquid milk into the dry milk powder.

10. State the merits of using skim milk powder in bread production.

11. Enumerate the advantages of using Milk solids in bread production.

12. Itemize the functions of NFDM.

CHAPTER 8

SUGAR

Sugar is the generic name for sweet-tasting, soluble carbohydrates, many of which are used in food.

There are various types of sugar derived from different sources. Simple sugars are called monosaccharides and include glucose (also known as dextrose), fructose, and galactose. The "table sugar" or "granulated sugar" more customarily used as food is sucrose, a disaccharide of glucose and fructose.

Sugar is used in prepared foods (e.g., Cookies and cakes) and it is added to some foods and beverages (e.g., coffee and tea). In the body, sucrose is hydrolyzed into the simple sugars fructose and glucose. Other disaccharides include maltose from malted grain, and lactose from milk. Longer chains of sugars are called oligosaccharides or polysaccharides. Some other chemical substances, such as glycerol and sugar alcohols may also have a sweet taste, but are not classified as sugars. Diet food substitutes for sugar include aspartame and sucralose, a chlorinated derivative of sucrose.

Sugar is a staple of baked goods, used in varying quantities in almost every variety. Breads and pancakes use a small amount of sugar, with around a few tablespoons, while dessert breads, cakes, pies and other desserts use large quantities of sugar, usually with more than a cup. Sugar has many purposes in baking, although it is possible to substitute it with artificial sweeteners.

Purpose

Sugar has many purposes in addition to adding sweetness to the baking recipes. Many recipes specify beating the sugar and a fat like butter, egg or oil to add air and fluffiness to the batter. This process helps increase the size of certain baked goods, such as cakes, when baked. The longer beat the two together, the fluffier the baked good will become. Sugar also holds water, which results in the baked good lasting longer. When baked, sugar often turns brown, changing the color of the recipe.

TYPES OF SUGAR IN BAKING

Granulated Sugar

What is granulated sugar? Granulated sugar is a refined sugar that is white in color and is the most common type of sugar used in baking. Granulated sugar has a slight coarseness to it but is still a very fine grain.

Best uses for granulated sugar: Granulated sugar is the sugar most commonly used in baking. Use it for almost any sweet baked good.

Brown Sugar (Light & Dark)

What is brown sugar?: Brown sugar is granulated sugar that has molasses added into it. Light brown sugar has a small amount of molasses added while dark brown sugar has a larger amount of molasses added into it. Molasses is even more hygroscopic in nature than plain granulated sugar so it keeps baked goods even more moist and adds chewiness.

It is worth noting here that molasses is actually a by-product of refining sugar. When sugar is refined, it is processed into smaller granules and the molasses is removed. This is then added back into the fine grain refined sugar to make brown sugar.

Best uses for brown sugar: Brown sugar should be used in baked goods where chewiness is desirable, such as for chocolate chip cookies. Use dark brown sugar for even more chewiness and more caramel flavors.

Powdered Sugar (AKA Confectioners' Sugar, Icing Sugar, or 10x Sugar)

What is powdered sugar?: Powdered sugar, which is also called confectioner's sugar, icing sugar, or 10x sugar, is a very finely ground white sugar. Because powdered sugar is so finely ground it is also combined with a bit of cornstarch, or other starch, to prevent it from clumping. Powdered sugar dissolves extremely quickly into baked goods, and because of its fine texture and the addition of cornstarch it can create very tender baked goods.

Best uses for powdered sugar: Powdered sugar works beautifully to make icings and frostings because it dissolves so easily. Powdered sugar also works well to dust over cakes or pastries as a simple decoration.

Superfine Sugar (Castor or Caster Sugar)

What is superfine sugar? Superfine sugar, also known as castor or caster sugar, is a more finely ground granulated sugar, though not as finely ground as powdered sugar. This type of sugar is popular for professional baking, and is very commonly used in the UK, because it more readily dissolves into batters and doughs. However, it is difficult to find and is typically pricier than granulated sugar.

Best uses for superfine (castor/caster) sugar: Superfine sugar is best used for baking uses when it is necessary to beat air into ingredients such as when whipping egg whites for a cake or meringue, whipped cream, or for frostings.

Muscovado Sugar

What is Muscovado Sugar?: Muscovado sugar is an unrefined cane sugar that is similar in texture to brown sugar due to the molasses naturally remaining in this type of sugar. Muscovado sugar has a very strong molasses flavor, more so dark than dark brown sugar, and is more moist than regular brown sugar. Muscovado sugar can be found in both light and dark varieties, similar to brown sugar.

Best uses for muscovado sugar: Muscovado sugar is best used for baked goods where a strong molasses flavor is desirable or complimentary of the other flavors such as for gingerbread, molasses cookies, or other baked goods with warming spices.

Sanding Sugar

What is sanding sugar?: Sanding sugar is a very coarse type of granulated sugar that is kept clear or sometimes colored a variety of colors.

Best uses for sanding sugar: For topping and decorating desserts.

Turbinado Sugar (Raw sugar or sugar in the raw)

What is turbinado sugar?

Turbinado sugar, also known as raw sugar or sugar in the raw, is a type of sugar that has been minimally processed. The texture of the sugar is very coarse, similar to the texture of sanding sugar, and is light brown in color. Almost all of the molasses is removed from this type of sugar so it is dry in texture but does have a hint of molasses flavor lingering.

Best uses for turbinado sugar in baking: Use it like sanding sugar, to top and decorate baking goods. Adds a crunchy texture.

Fig. 7. Granulated Sugar

Fig. 8. Brown Sugar (light)

Fig.9. Brown Sugar (Dark)

Fig. 10. Icing Sugar

Fig. 11. Castor or Caster Sugar

Fig. 12. Muscovado Sugar

Fig. 13. Sanding Sugar　　　**Fig 14. Turbinado Sugar**

Fig. 15. Pearl Sugar

Pearl Sugar (Nib Sugar)

What is pearl sugar?: Pearl sugar, also called nib sugar, is a type of specialty sugar that is made by compressing granulated sugar into large hunks of sugar. This type of sugar is only used for very specific baking purposes as it does not dissolve into baked goods.

Best uses for pearl sugar: Pearl sugar is most commonly used to make Belgian Liege Waffles but is also used as a very coarse topping on a variety of pastries.

Making Sugar Substitutions in Baking

Because sugar does have so much effect on the structure and texture of a baked good, making substitutions with sugar should be done with caution. It is also important to note that each type of sugar has varying density levels so it can be substituted one for another it is important to do so by weight instead of by volume.

Swapping granulated sugar and brown sugars: In most cookies, brownies, and bars, it is typically fairly safe to swap granulated sugar and brown sugar. Granulated sugar will make baked goods more crisp while brown sugar will create chewiness and a more moist texture.

Using powdered sugar in place of granulated sugar: This substitution should only be done in baked goods that are fairly forgiving such as cookies, brownies, bars, muffins, and quick breads. Make sure that measuring by weight as the amount of volume needed of granulated sugar is significantly less than powdered sugar.

Swapping brown sugar and muscovado sugar: These two sugars function very similarly and can typically be swapped without issues. Keep in mind that muscovado sugar will add a much more intense molasses flavor to the baked goods.

Properties of sugar in baking

Hydrolysis: Compound sugars like sucrose are split into their component sugars by specific enzymes or acids. Maltose and sucrose are hydrolyzed by the enzymes maltose and invertase, respectively. Both these enzymes are present in bakers' yeast. Those reactions take place in the dough before the sugars are fermented. Sucrose is converted into two simple sugars, fructose and dextrose, so rapidly that the hydrolysis is complete a few minutes after mixing and it is so thorough that practically no sucrose is detectable in the finished bread. In contrast, almost all of the original lactose content remains in bread because yeast does not have an enzyme to hydrolyze lactose.

Yeast Fermentation: Glucose, fructose, sucrose and maltose are readily fermented by bakers' yeast to produce carbon dioxide and alcohol as principal end products. Lactose is not fermentable because baker's yeast lacks the enzyme which could split this compound sugar.

Residual sugars: About two per cent of the sugars added, based on flour, are used up during the bread fermentation. The remaining sugars which are present in bread are called residual sugars. Therefore, the higher the percentage of sugar used in the formula, the higher is the amount of residual sugars.

Sweetness and flavor: Since there is no physical or chemical test for sweetness it must, therefore, be related to taste.

Hygroscopicity and Hydration: Hygroscopicity is the ability of a substance to absorb moisture and retain it. Some sugars are more hygroscopic than others.

Heat Susceptibility: When sugars are heated, molecules combine to form colored substances called Caramel. Sugars vary in their heat sensitivity, i.e the temperature at which they begin to caramelize. Fructose, maltose and dextrose are more sensitive and lactose and sucrose are the least. By lowering the pH of the sugar solution, fructose and dextrose become less sensitive.

Browning reaction: Reducing sugars, when heated with proteins, react to form dark compounds called melanoidins.

Solubility and Crystallization: The difference in solubility of sugars can be used to control crystallization in products that require higher amounts of sugar.

Softening: The tenderizing action of sugars in baked products with the resultant improvement in texture, volume and symmetry may indirectly be attributed to the ability of sugar to hold water.

Functions of sugar in baking

Sugar harvested from cane sugar is chemically identical to sugar harvested from sugar beets. The two are not easy to tell apart from each other and likely, have purchased both. Sugars perform several functions in bread and other bakery products. The functions of sugars in baking are listed below:

- Sugars are the source of energy for yeast activity.

- Sugars are either provided by starch hydrolysis or by direct addition to the formula.

- The flavor of the bread is improved and the crust color is darkened by the addition of sugar.

- The texture and grain become smoother and finer with added sugar. It may be related to the action of sugars on delaying the gelatinization of starch and the denaturalization of protein.

- Sugar can be heated into a syrup or caramelized and used to make intricate sugar decorations.

- Sugar stabilizes the egg foam.

- Sugar holds onto moisture, baked goods made with sugar do not stale as quickly as baked goods made without sugar.

- Sugar holds onto moisture; it keeps baked goods tender for a longer period of time.

- Sugar also serves to help reduce gluten development and tenderize baked goods.

- Sugar in baking is that it adds sweetness and flavor.

- Sugar serves to help leaven baked goods in a variety of ways.

- Sugar essentially serves as a cushion between the bubbles which stabilizes the egg foam.

- Sugar syrup or caramelized sugar is used to make intricate sugar decorations.

In text Questions

1. Unrefined can sugar is called as _____ because of its intense molasses flavor and aroma.

2. Sugars are the source of energy for _____ activity.

3. _____ essentially serves as a cushion between the bubbles which stabilizes the egg foam.

4. What are the different types of sugars?

5. Write the properties of sugar.

6. Why do we need sugar in baking and discuss its functional role in baking?

CHAPTER 9

SALT

Salt occupies a special position as an ingredient to enhance the taste palatability and flavor of foods. It is also a preservative of a variety of food products when used in large amounts. Cooking salt is almost entirely sodium chloride with other salts such as magnesium and calcium chlorides and magnesium, calcium and sodium sulphates constituting about 2-25% other trace elements are also found to be present in the cooking salt. Salt plays a role in tightening the gluten structure and adding strength to the dough. It helps the loaf to hold on to the carbon dioxide gas that is formed during fermentation, supporting good volume. In bread baking, salt controls yeast growth and has a tightening the gluten structure and strengthening effect on the gluten in the dough. Also helps the loaf to hold on to the carbon dioxide gas that is formed during fermentation, supporting good volume. Salt slows down fermentation and enzyme activity in the dough. In pastry-making, it helps cut the oily mouth feel of buttery doughs and encourages browning. Next to its role in boosting the flavor of bread, salt plays a role in and adding strength to the dough.

Functions of salt in baking

- Salt is an essential ingredient for most baked foods performs functions in baking that cannot be duplicated by any other ingredients.

- The ordinary granulated salt being relatively slow dissolving and contains higher levels of impurities (Copper, Iron & other metals), increases mixing time and accelerates gluten formation and hence toughen the dough through a direct action on the dough proteins or due to an inhibitory action of proteolytic enzymes and hence it is suitable for bread making.

- Helps to control the yeast activity in bakery products.

- Prevents the formation and growth of undesirable bacteria in bakery products.

- Improves the natural flavor of the products.

- Improves WAP.

- Improves the texture and grain of the products.

- The crust color of the product is enhanced by lowering the caramelization temperature of the sugar.

- Helps to cut down the excessive sweetness in bakery products.

- Keeps baked goods, fresh and moist for a longer period of time

- Controls the production of unwanted acids in dough.

- A pinch of salt always improves the taste and flavor.

- Lesser tightening effect on the dough is preferred in cookie and biscuit preparation and improves gas retention power.

- Not only salt imparts a flavor that makes the taste of a product good, but it also acts to accentuate the flavor of other major and minor ingredients, for example, sweetness of sugar is emphasized by the contrasting taste of salt.

- Salt enhances the natural flavor of other ingredients used in cake making and thus improves the flavor of cakes.

- It improves the crust color of the cake by lowering caramelization of sugar.

- Salt helps in cutting down excessive sweetness in cake.

In text Questions

1. _____ is an ingredient to enhance the taste palatability and flavor of foods.

2. _____ is a preservative of a variety of food products when used in large amounts.

3. Cooking salt is otherwise called as _____.

4. Salt plays a role in tightening the_____ structure and adding strength to the dough.

5. What are the uses of salt and explain how it aids in making bread?

6. Enumerate the functions of salt in baking.

CHAPTER 10

WATER

The material that provides moistness to the products is known as "moistening agent" in our bakery we are using many moistening agents to prepare bakery and confectionery products.

Water plays an important role in bakery and confectionery production. The chemical formula of water is H_2O. Fresh water is divided into three categories according to the source.

They are

1.   Natural Water

2.   Ground Water

3.   Surface Water

Water then it is classified into different categories depending on the presence of carbonic matter and minerals. They are

1.   Hard Water

2.   Soft Water

3.   Acidic Water

4.   Alkaline Water

5.   Saline Water

6.   Drinking Water

Natural water

Is obtained from snow and rain, ground water is obtained from springs, shallow and deep wells and infiltration galleries. Surface water is obtained on the ground surface like rivers, streams, lakes, sea and other natural reservoirs.

Hard water

It contains more dissolved minerals. The hardness may be of varying degree and is measured in terms of ppm [parts per million]

Soft water

Contains very low dissolved mineral. Acidic water has pH less than 7. This rarely found in nature. The pH of alkaline water is more than 7. This kind of water is generally found in certain regions where there is considerable alkali in the soil. Saline water contains sufficient sodium chloride, which can be sensed by the taste.

Drinking Water

That means for either drinking or for food preparation should be free from organic matter and bacteria.

For production, bakers prefer to use medium hard water for bread production, because the limited quantities of minerals present in it has a beneficial effect on gas production, as the yeast requires minerals for vigorous fermentation.

Functions of Water

• It gives moisture to the products.

• It combines all the dry ingredients together.

• It helps to distribute the raw material equally in the batter.

- It builds the structure of the baked products.

- It controls the batter and dough temperature.

- It controls the consistency of the dough and batter and affects the volume and texture of cakes.

- It helps to develop the gluten, when the flour is mixed with water and kneaded, it forms flour gluten.

- It helps to release CO_2 gas [example baking] and formation of vapor pressure and it gives volume to the products.

- It improves the keeping quality.

- Water is converted into vapor during baking, which requires mores space and thus leavens the products.

In text questions

1. What are the different sources of water?

2. Describe the types of fresh water in brief.

3. List down the different functions of water in bakery products.

4. How water helps in baking process?

CHAPTER 11

FLAVORING AGENTS

Flavor may be defined as the sensation of smell and taste mingled. The flavor is an important ingredient in a sweet good formula. The flavor is really the ingredient which helps the baker to add a uniqueness to the product and it is used in comparatively small quantities in the baked products. A baker can add a variety of tastes to the baked products by choosing fresh and high quality spices. The general accepted components of taste are: Sweetness, sourness, saltiness and bitterness.

Cinnamon

This is the most widely used spice. This plant is grown in China, Indo-China and Indonesia. Cinnamon is used in making of cakes, cookies, pies and custard fillings. It is also widely used in various varieties of Danish pastries.

Cardamom Seed

Cardamom is mostly grown in India and Sri Lanka. Cardamom is mostly used in Nankhatai, Cookies, Danish pastries, and in the fillings of Éclairs and Pies.

Ginger

Ginger is the root of a tuberous plant mostly grown in India, Jamaica and Africa. Ginger is grown in Cochin is considered to be of the best grade for baking.

Cloves

Cloves are the dried, unopened buds of an evergreen tree grown in Indonesia, Zinzibar, Madagascar.

Nutmeg

Nutmeg is the seed of evergreen tree grown particularly in the Molucca Islands, Indonesia and the West Indies. This Spice is used mainly in doughnut and pastry crusts.

Mace

It is an aromatic spice consisting of the dried external fibrous covering of a nutmeg. This is mainly used in sponge and pound cakes, cream fillings of éclairs.

Poppy Seeds

Poppy is raised in Turkey, Iran, India, Netherlands, Russia and Poland. There are two varieties of poppy-creamish white and blue poppy. Blue poppy seed is considered the best as it is used for grain as well as for flavor. The seeds are mainly for sprinkling on tops of variety breads and rolls.

Caraway Seeds

Caraway seed is the fruit of the tree belonging to the parsley family. It is grown in Europe, particularly in the Netherlands and Poland. In the baked products it can be used as whole as well as ground. This is a must in the making of rye bread and it is also used in rich fermented cookies-known as Surti Butters'–most popular variety of the baked products on West Coast of India.

Sesame seeds

Sesame seed is a small honey colored seed grown mainly in Turkey and India. It is used for topping the bread and rolls and when baked imparts delicious, roasted nut flavor to the crust.

Allspice

It is a fruit of the pimento tree which is grown in Jamaica, Mexico and other parts of central and South America. These fruits are dried and then ground. Their flavor resembles the blended flavor of nutmeg, clove and cinnamon. Its uses in baking are for making fruit cakes and cookies and also in filling of pies.

There are various sources through which the baked product can acquire its unique bakery flavor.

It can acquire the flavor during the processing of the product, i.e during baking, fermentation etc.

a. **Fermentation:** The total fermentation time has a profound influence upon the end flavor of the baked product due to the biological changes that take place during the fermentation. Breads made from sour dough or overnight sponge have a different flavor from the breads made from short sponges and straight dough process.

b. **Baking:** The process of baking brings about two important changes which add flavor to the product:

    i.   Brown reaction

    ii. Caramelization

Flavor Additives

These additives can be divided into three groups:

Natural, synthetic and imitation (with unlimited combination of all three)

1. **Natural**

    a. Basic ingredients added to the formula: forms of sugar and syrups i.e. honey, molasses malt syrup etc., ground fresh fruit, cocoa, chocolate etc.

    b. The essential oil of citrus fruits such as oil of lemon and oil of orange and vanilla extract

2. **Synthetic:** The quantities of flavors present in the fresh fruit are very small. If the flavor of the fresh fruit was to be used singly alone in the formula, large quantities of ground, sliced fruit will be necessary to bring about the desired level of flavor. This will not only unbalance the formula, but will make it impracticable. If this natural flavor is fortified with synthetic flavor it will have more taste appeal than the use of natural flavors alone.

3. **Imitation:** The imitation of natural flavors is rarely used alone, but are blended with fruit juices and essential oils to give a better result. Imitation flavors are not found in nature but used to reproduce the natural flavor.

Classification of Flavors

i. **Non-alcoholic flavor:** These are prepared by dissolving ingredients in Glycerine propylene, glycol or vegetable oil. These help to retain the flavor during baking by reducing vaporization.

ii. **Alcoholic extract:** These flavors dissolve in ethyl alcohol. They are too volatile to use during baking, but are very suitable for icing and fillings which do not undergo baking.

iii. **Emulsifiers:** These flavoring oils are dispersed in a gum solution which help to obtain an even distribution through batter and dough and also maintaining stability during baking.

iv. **Powdered flavorings:** These are prepared by emulsifying components in heavy gum/water solution, then spray dried to form powders.

Anything can cause off flavors in the baked products and some of the causes are:

1. Inferior Ingredients: These are the prime causes for end products having off flavors-musty flour, moldy cake or bread crumbs, eggs which are not fresh, shortening low in anti-oxidant, spoiled milk etc.

2. Unbalanced Formula: The formula should be balanced in such a way that the total effect of this should have an appealing blend of the foundation flavors of all the ingredients included in the formula.

3. Inferior flavorings: Not only the inferior flavorings but too much of the flavoring will also give the product an off flavor.

4. Wrong pH: This can happen if an excess of soda is added in the formula or if there is too much of acid produced in the product. Excess of Soda in the formula will have the undermentioned defects in the end products: excessive caramelization, crumbliness and poor texture, dryness and discoloration of the crumb and soapy taste. Each of these faults individually or in combination has a direct effect on the flavor and aroma of the finished product.

5. Faulty baking as well as faulty processing: These will cause either excess or lack of flavor.

6. Unclean pans: If the pans used for baking are not thoroughly cleaned of residual material, it may cause an off flavor in the product. Wrong pan grease and improper storage of cleaned and greased pan should be avoided.

7. Cheap and wrong type of packaging and wrapping material: These will affect the flavor of the finished product.

8. Poor ventilation and lack of proper air circulation: Within the bake house these conditions may cause off flavor.

9. Improper storage of finished products: This should be totally avoided. The finished products having separate flavors should not be stacked together at the time of the cooling and packaging.

10. Dirty and defective transportation: This will also cause off flavor in the finished product.

Storage of Spices

The volatile oils in the spices contain the aroma and flavor of the spices. In order to retain the strength of these oils, spices should be stored in an airtight container. Storage room should be dry, cool and airy.

Nuts and Fruits

A variety of dried and preserved fruits and nuts can be used in baked products to produce different types of flavors and finishes.

It is usually necessary to wash dried fruits before use with a liberal amount of water and swirled around for about one minute. Care must be taken so that the fruit does not absorb too much water and become soft. If fruit absorbs too much water it will break down during mixing and discolor the dough. The flavor also diminishes if the fruit is soaked too long. After washing the fruits should be drained in a sieve. After draining the fruits should be carefully picked over by spreading the fruit on a dry cloth to remove the excess moisture. The fruit should always be added last to ensure even distribution throughout the batter / dough with minimum damage.

Dried fruits

Among all the dried fruits, the products of different types of grapevines take the foremost place in the confectionery.

Currants

Currants are the dried form of small black grapes. Good quality currants should be bold, fleshy and clean of even size and blue black in color. The currant should not contain red shriveled berries which due to their extra acidity spoil the flavor of cakes. Currants prior to their use should be soaked in boiling water for 2 minutes and then dried to get rid of the grit, stalks and stones.

Sultanas: Sultanas are made out of seedless yellow grapes.

Raisins

Ripe grapes are converted into raisins and sultanas in different ways. In the case of raisins, the bunches of grapes are partly dried by twisting the stalks while still on the vine and then are finished off in open sheds.

Dates

Dates are the dried fruits of Iraq and North African palms. They are very sweet and rich in sugar. Dates should always be soaked in about half of their weight of water for an hour or more until they are soft.

Sugar preserved Fruits and Peels

The skin or peel of the citrus fruits such as lemons, oranges and citrons is abundantly used in fruit cakes. Thick rind fruits are the best for the preparation of peels. These fruits are cut across the middle and the pulp is removed. The halves, known as caps, are soaked in brine for several days to remove the acid taste of the rind.

Cherries are available as glace and as crystallized.

Glace cherries

Good quality fruit is bleached in a solution of water, calcium carbonate and Sulphur dioxide until the fruit is colorless. The fruit is stoned and washed. Cherries are then soaked for a few minutes to soften the skin and flesh. After draining, cherries are immersed in weak syrup, which is colored, usually red, but also green or yellow. The syrup strength is increased daily by boiling over a period of 9 days.

Crystallized Cherries

The crystallized cherries are made by draining the preserved glazed cherries and rolling them in fine castor sugar.

Crystallized fruits

Crystallized fruits are mostly used in the decoration of rich fruit and other cakes. Before using, the crystalline sugar from the fruits is washed off and the fruits are cut into the desired shapes and placed on the cake. After baking the fruits can be washed with a good syrup to enhance the brightness of the fruit. Pineapple, peaches, apricots, plums and pears are generally used to make crystallized fruits.

Angelica

Angelica is a large green plant of which only the stem is used. It is preserved in a similar way to cherries, using green syrup. Apart from the bright green color it has an aromatic flavor.

Ginger Root

Only the tuberous root of the plant is used. It is washed well and boiled in a weak sugar solution until soft. The sugar strength of the syrup is to be gradually increased as in the candying process of citrus peels. It is stored in syrup as root, chips or crushed.

Crystallized Flowers

Rose petals are laid out on wires and suitably colored syrup is allowed to drip on to them. When thoroughly saturated, the petals are dried over gentle heat. They can be used as a decoration, owing to the colors but also have a scented flavor of the original flower.

Nuts

Nuts offer various flavor, texture, bite and appearance in baked products-especially in the products of cookies. There are several nuts available in various shapes and all have a high food value. Most of the nuts are expensive, which restricts their use only in specialty items.

Almonds

Among the nuts the almonds have a richness and fineness of flavor, but due to their high cost are used in specialty items and for decorations. There are two types of almonds: the sweet and the bitter. The bitter almond is used in the preparation of essential oil of almond. Bitter almond is not suitable for eating and is used to boost up the flavor by blending with sweet almonds. In confectionery, almonds are used either whole or split or ground or a combination thereof according to the type of desired end products.

Walnuts

Walnuts have a strong flavor and due to high fat content have a tendency to go rancid after a long storage.

Pistachio Nuts

Pistachio Nuts are sparingly used in the baked products because of their high price. They are about ½ ‾ in length and have a purplish brown skin, which can be removed by blanching.

Cashewnuts

They have a bland flavor. They are largely used for decorative purposes.

Groundnuts (Monkey Nuts or Peanuts)

They contain about 40% of oil, which is used in making vegetable fats. Because of price advantage, groundnuts are generally used in various cookie recipes.

Coconuts

The white flesh known as copra is removed from the shell of the nut and is dried either in the sun or in the shade. A better color is produced by shade drying. When dried, the copra is cut according to confectioner's requirements such as shredded, coarse, medium or fine desiccated coconut. Cut coconut can also be colored by mixing well with a liquid color and then drying off the excess moisture. Owing to high oil content, coconut is liable to develop rancidity. Coconut has a tendency to be contaminated with salmonella bacteria which is harmful for health.

Functions of flavoring agents in baking

- The flavor is a very important aspect of quality product.
- A flavoring agent's action should not be impaired due to heat or storage.
- Cheap flavors often break down under the influence of heat giving off flavor to the product.
- A flavoring material should be added after measuring, the excess will spoil the gastronomic appeal of the product.

In text questions

1. Excess or lack of flavor is due to faulty _____ and _____.

2. Walnuts have a strong flavor and due to high fat content have a tendency to go _____ after a long storage.

3. _____ are the dried form of small black grapes.

4. Coconut has a tendency to be contaminated with _____ bacteria which is harmful for health.

5. Enlist the flavoring agents used in baking. Brief the agents used in baking.

6. State the role of spices in baking

7. How nuts and dried fruits used in bakery products?

8. What is the function of flavoring agents in baking?

CHAPTER 12

LEAVENING AGENTS

A leaven often called a leavening agent (and also known as a raising agent), is any one of a number of substances used in doughs and batters that produce a foaming action (gas bubbles) that lightens and softens the mixture. An alternative or supplement to leavening agents is mechanical action by which air is incorporated. Leavening agents can be biological or synthetic chemical compounds. The gas produced is often carbon dioxide, or occasionally hydrogen.

When a dough or batter is mixed, the starch in the flour and the water in the dough form a matrix (often supported further by proteins like gluten or polysaccharides, such as pentosans or xanthan gum). Then the starch gelatinizes and sets, leaving gas bubbles that remain.

The leavening of bakery products could be brought about by the following four general ways:

1. Leavening by mechanical way (aeration)
2. Leavening by biological way
3. Leavening by water vapor
4. Leavening by chemical way

During mechanical aeration, there must be some ingredients in the mix that will hold the air bubbles and not allow them to escape. This is normally brought about by protein substances such as egg, egg white (albumen) or gelatin. Examples of products of this method are the production of marshmallow and icing (royal).

Another method of mechanical aeration is in the layers of dough with an insulating material in between. In such cases, a dough is used with well-developed gluten present. When the biscuit enters the oven the water in the dough layer is converted to steam and expands, lifting the layers. Examples of such leavened products are crackers and puff biscuits.

There are two commonly followed mechanical ways by which the air is incorporated into dough or batter, viz., (a) creaming, and (b) beating/whipping.

a) **Creaming:** During the creaming process the air is entrapped into the shortening, which expands when heated during the process. This gives volume to the bakery product. The pound cake is an example of this process, wherein none of the chemical leaveners are used.

b) **Beating/Whipping:** When egg whites are beaten or whipped they become fluffy and foamy because of the whipped-in air. This air incorporated during whipping of eggs, expands while the batter is being baked and causes the cake to rise. The sponge cake is example of cakes leavened by this manner.

Chemical method [by releasing of CO_2]

Some of the commonly used chemical leaveners are:

a) Baking soda

b) Ammonium bi-carbonate

c) Baking powder

a) **Baking Soda:** The chemical name of baking soda is sodium bicarbonate and has the chemical formula $NaHCO_3$. It is also known as 'bicarbonate of soda'. During the baking process, it liberates carbon-di-oxide, a leavening gas. The chemical reaction of gas formation is as follows:

$$2 \text{ NaHCO}_3 \longrightarrow \text{Na}_2\text{CO}_3 + \text{CO}_2 + \text{H}_2\text{O}$$

Baking soda also liberates carbon-di-oxide gas when it is mixed with acidic substances like sour milk, buttermilk. The washing soda being alkaline increase the pH of the cake batter and thus enhances the rate of caramelization of sugar, giving darker crust color.

- The liberation of carbon di-oxide gas from pure solutions of baking soda is slow, especially near room temperature. When baking soda is added to dough or batter, gas liberates at least initially. In the absence of added acids, the dough pH quickly becomes alkaline and gas production decreases. The popularity of sodium bicarbonate as a gas source is based on its low cost, lack of toxicity, ease of handling and easy availability. When heated it releases sodium carbonate water and CO_2.

- This sodium carbonate gives an unpleasant smell to the product and color of the cake will be brown.

b) **Ammonium Bicarbonate:** Ammonium bicarbonate is used rather extensively in cookies and in bakery products that are baked almost to dryness. During the baking process, it decomposes completely into gases like ammonia, carbon-di-oxide and water vapor.

In other words, ammonium bicarbonate is called volatile salt. The chemical leavening reaction of this salt could be written as under:

$$\text{NH}_4\text{HCO}_3 \longrightarrow \text{NH}_3 + \text{CO}_2 + \text{H}_2\text{O}$$

Use of slight excess can completely break down the structure of the biscuit or cookies.

c) **Baking powder:** Baking powder is a chemical leavening agent produced by blending a water soluble sodium bicarbonate (baking soda). One or more acid reacting ingredients with or without any filler, such as starch, calcium carbonate or flour.

Baking Powder = Sodium Bicarbonate + One or more acid reacting material + Inert filler (starch, calcium carbonate or flour) (25-30% variable)

Since the acid-alkali reaction (taking place between constituents of baking powder) does not take place until both soda and acid ingredients are in solution, one of the acid ingredients should not be soluble, except in warm or hot water. Sodium bicarbonate is an alkaline salt which is soluble in cold water. It follows, then, that a suitable acid ingredient should be used which does not dissolve during baking operation.

A chemical which is mixed with the recipes to increase the volume of products are called chemical agents.

Fast Action

It releases most of CO_2 gas during bench operation and very little gas during baking.

Slow Action

It does not release much of the CO_2 gas during bench operation, but all the gas it released when it comes in contact with heat.

Double Action

It is a combination of the fast and slow acting types of baking powder is most widely used by bakers.

Ammonium Bicarbonate or Bicarbonate

- When heated, it release ammonium carbon dioxide and water
- Ammonium bicarbonate is used in cookies, crackers and similar products
- If the gas does not escape completely the taste and odor of ammonia remains in the bakery products.

The several acid ingredients used in the preparation of various types of baking powders are:

1. Tartaric acid

2. Citric acid

3. Cream of tartar

4. Acid calcium phosphate

5. Sodium acid pyrophosphate

6. Monosodium phosphate

7. Glucono-delta lactone

The chemical reaction between sodium bicarbonate and some of the selected acid ingredients is shown below:

1. Tartaric acid

2. Cream of tartar

Baking powder is a solid mixture that is used as a chemical leavening agent in baked goods. It can be composed of a number of materials, but usually contains baking soda (sodium bicarbonate or $NaHCO_3$), cream of tartar (potassium bitartrate or $C_4H_5KO_6$), and cornstarch. A base, an acid, and a filler respectively. And secondly, none of these components of baking powder is found as residue at the end of the reaction, thus, there will not be much effect on the pH of the batter/dough and saponification reaction too cannot take place.

If sodium bicarbonate alone is heated, only half the total amount of carbon-di-oxide is released. If an acid ingredient is used to react with the sodium bicarbonate (an alkaline salt), all the carbon-di-oxide is released and there is no noticeable action on the gluten or on the crumb color. In this way, there is a considerable yield of carbon-di-oxide for aeration. Secondly, it will also neutralize the washing soda (Na CO which forms on decomposition of baking soda).

Biological leavening agents

Yeast is actually a member of the fungus family and is a living organism in the air all around us. Yeast consists of microscopic, unicellular plants which are capable of rapid multiplication when conditions are favourable and which obtain energy by breaking down sugars to carbon dioxide and alcohol. This process is known fermentation and is brought about by the enzyme zymase found in yeast. Yeast also produces enzymes which are able to split disaccharide sugars.

Baker's yeast (like baking powder and baking soda) is used to leaven baked goods such as breads and cakes. Baking powder and baking soda react chemically to yield the carbon dioxide that produces the baked products rise. Yeast, however, does not produce a chemical reaction. Instead, the carbon dioxide, it produces is the result of the yeast, exactly, feeding on the dough.

Yeast comes in two forms: (1) Fresh Yeast (also called Compressed Cakes) and (2) Dry Yeast (also called Dehydrated Granules).

Fresh yeast: Fresh yeast is soft and moist mixture of yeast plants and starch and is mainly used by professionals. The yeast remains active and will grow and multiply rapidly when added to dough. It has to be kept at refrigerated or frozen temperatures, as it is highly perishable. It keeps well only for a few days. Fresh yeast needs to be proofed before using.

Dry yeast: Dry yeast is a fresh yeast that has been pressed and dried until the moisture content makes the yeast in an inactive state (until mixed with warm water). Dry yeast has a much longer shelf life than fresh yeast and does not need to be refrigerated unless opened. Once opened, the dry yeast needs to be stored in the refrigerator away from moisture, heat, and light because it deteriorates rapidly when it is exposed to air.

There are two types of dry yeast: (Regular) Active Dry Yeast and Rapid-Rise Yeast. Though there are some slight variations in shape and nutrients, Rapid-Rise Yeast is the same as Instant Yeast and Bread Machine Yeast.

These two types of dry yeast can be used interchangeably, with some limitations. Though Bread Machine Yeast is faster-rising and is specially formulated for bread machines, as its texture is finely granulated to hydrate easily when combined with flour. Active Dry Yeast may also be used in bread machines though it but may not yield completely equal results. The advantage of the Rapid-Rise Yeast is the rising time is half that of the Active Dry and it only needs one rising. Though this is an advantage, some flavor and texture by speeding up the rising process as the yeast does not have time to develop its own flavor. Also, Rapid-Rise Yeast is a little more powerful than Active Dry Yeast and can be mixed in with dry ingredients directly.

The yeast used in the preparation of the fermented bakery products (like bread, bread rolls, sweet doughs, crackers etc.) does the job of leavening by biological way. Here carbon dioxide gas is generated by fermentation. The one best adopted for the leavening of bakery dough is baker's yeast. Sugars such as glucose and fructose are substrates, which are transformed into carbon-di-oxide and ethyl alcohol by fermentation. A simplified equation describing this fermentation reaction could be written as:

$$C_6 H_{12} O_6 \longrightarrow 2C_2 H_5 OH + 2 CO_2$$

Glucose/Fructose   Ethyl alcohol   Carbon-di-oxide

This carbon-di-oxide is responsible for the leavening of bakery products. The advantages of yeast leavening, as opposed to chemical leavening, are that it can contribute a characteristic taste and aroma and the evolution of gas can continue over a much longer period of time.

Water used in bakery formulations either in the visible form (pure water) or invisible form (as moisture in ingredients like milk, egg, syrups, etc) changes to water vapor as the temperature of a cake batter or bread dough rises during baking operation. It is also true that unless the consistency of batter or dough is adjusted properly, the chemical leaveners alone cannot do their work efficiently. Because the stiff dough or batter does not rise much and the loose dough or batter cannot retain the expanding gases thus causes collapse resulting in lower volume.

Microorganisms that release carbon dioxide as part their lifecycle can be used to leaven products. Varieties of yeast are most often used. Yeast leaves behind waste byproducts that contribute to the distinctive flavor of yeast breads. In sour dough breads the flavor is further enhanced by various lactic or acetic acid bacteria. Leavening with yeast has been often a slower process requiring a lengthier proofing.

Yeast can also be used to make carbonated beverages like beer, which can then be used as leavening. Some typical biological leaveners are:

- Beer (unpasteurized—live yeast)
- Ginger beer
- Butter Milk
- Sour dough
- Yeast
- Yogurt

Some of the advantages of the leavening of bakery products are as under:

- It increases the volume of the bakery products.
- The leavened products being light and porous are easily chewed and digested.

- Leavened bakery products are more palatable and appetizing than those made without leavening, which may be due to uniformity of cell structure, brightness of crumb color, softness of texture, etc.

- Baked products so made are light, and therefore easily chewed.

Because baked products made with leavening have an open or more porous grain than unleavened products, the digestive juices come in contact with the food more readily and digestion is greatly facilitated. Baked products made with leavening agents are more palatable and appetizing than those made without leavening.

Functions of yeast

The primary function of yeast is to leaven the dough or to make it rise-and to produce a porous product.

Production of carbon dioxide: Carbon dioxide is produced by the yeast as a result of the breakdown of fermentable sugars in the dough. The development of carbon dioxide causes expansion of the dough as it is trapped within the protein matrix of the dough.

Causes dough maturation: This is accomplished by the chemical reaction of yeast produced alcohols and acids on protein of the flour and by the physical stretching of the protein by carbon dioxide gas. This results in the light, airy physical structure associated with yeast leavened products. Phytate in the dough will decrease the uptake of minerals due to complexing.

Development of fermentation flavor: Yeast imparts the characteristic flavor of bread and other yeast leavened products. During dough fermentation, yeast produce many secondary metabolites such as ketones, higher alcohols, organic acids, aldehydes and esters. Some of these, alcohols for example, escape during baking. Others react with each other and with other compounds found in the dough to form new and more complex flavor compounds. These reactions occur primarily in the crust and the resultant flavor diffuses into the crumb of the baked bread.

In text questions

1. _____ is the leavening agent used in biscuit making only.

2. _____ is the biological leavening agent.

3. _____ is known as giving volume to any bakery product.

4. Baking powder is the combination of _____ and _____.

5. Creaming is the example of _____ leavening.

6. Baking soda is _____ by pH value.

7. _____ leavening is used to give volume in puff pastry.

8. Pound cake requires _____ leavening.

9. Excess of baking soda results into _____ crumbs.

10. Leavening agents are also known as _____ agent.

11. When the volume _____ it is called leavening.

12. Through leavening the texture becomes _____ in baked products.

13. Soda and acid are termed ————–— ingredients in baking powder.

14. Name the different leavening agents. Differentiate between baking powder and baking soda.

15. Define creaming method.

16. What are leavening agents? Explain chemical leavening agent.

17. What reaction takes place while using the baking powder?

18. What are the main ingredients of baking powder? What are the acid ingredients used in the preparation of baking powder?

19. Why is baking powder used more frequently as compared to baking soda and Ammonia bicarbonate?

20. What is yeast? Classify the types of dry yeast. What is biological leavening?

21. How will you use yeast? How yeast helps in giving volume to the bakery product?

22. What is the importance of leavening/aerating?

23. What is the result in the properties of a leavened product? Explain.

24. Define leavening agents and its uses in bakery.

25. Discuss the leavening action of yeast on bread dough.

CHAPTER 13

BAKING PROCESS

Basic concepts of baking process

- Fats melt
- Gases form and expand
- Microorganisms die
- Sugar dissolves
- Egg, milk, and gluten proteins coagulate
- Starches gelatinize or solidify
- Liquids evaporate
- Caramelization and Maillard browning occur on crust
- Enzymes are denatured
- Changes occur to nutrients
- Pectin breaks down

Batch or continuous dough mixing

A batch process is a process in which one batch is made and finished before the other is started. In a small scale bakery, most processes will be in batches. Kneading the dough will be done in a bowl with a mixer blade. Only when all the dough is over, the bowl will be emptied and the next batch is started. The same for the oven, a rack of dough is put in the oven, only when this batch is finished will a new batch be present.

A continuous process on the other hand goes on continuously. In other words, no need to wait until a batch is completed before beginning the new process, instead, keep on doing the one process. Again, when looking at a bakery, in a large scale bakery, the oven is run continuously. Bread is continuously going in and out. This is done by using a belt which runs through the oven, this keeps on running all the time.

Dough Make-Up

The function of dough make-up is to change the fermented bulk dough into accurately sealed and moulded dough piece, when baked after proofing, it produces the desired finished product. Dough make-up includes (a) scaling (b) rounding (c) intermediary proofing (d) moulding.

Scaling or dividing

The dough is divided into separate pieces of fixed even weight and size. The weight of the dough to be taken depends on the final weight of the bread required. Generally, 12% extra dough weight is taken to compensate for the loss. Dividing should be done within the shortest period in order to make sure the uniform weight. If there is a delay in dividing, corrective steps should be applied either by degassing the dough or increasing the size of the dough. The degassers are essentially dough pumps which feed the dough into the hopper and in the process remove most of the gas. The advantages of using degassers are: (i) more uniform scaling (ii) uniform pan flows and (iii) uniform grain and texture of bread.

Rounding

When the dough piece leaves the divider, it is irregular in shape with sticky cut surfaces from which the gas can readily diffuse. The function of the rounder is to impart a new unbroken surface dough that will retain the gas as well as reduce the stickiness, thereby increasing its handling. There are of two types of rounder - umbrella and bowl type.

Intermediate proofing

When the dough piece leaves the rounder, it is somewhat degassed as a result of the mechanical, it received in that machine and in the divider. The dough lacks extensibility and tears easily. It is rubbery and will not mould easily. To restore more flexible, pliable structure which will respond well to the manipulation of moulder, it is necessary to let the dough piece rest while fermentation proceeds. Intermediate proofer contains a number of trays. The dough pieces are deposited in the tray. Average time at this period ranges from 5 to 20 min.

Moulding

Then the moulder collects pieces of dough from the intermediate proofer and shapes them into cylinders ready to be placed in the pans. Moulding involves three separate steps; (i) sheeting; (ii) curling; and (iii) scaling. Sheeter degasses the dough and sheeted dough can be easily manipulated in the later stages of moulding. Sheeting is accomplished by passing the dough through 2 or 3 sets of closely spaced rolls that progressively flatten and degas the dough. The sheeted dough piece next enters the curling section. A belt conveyor under a flexible woven mash chain that rolls into a cylindrical form carries the sheeted dough. The rolling operation should produce a relatively tight curl that will avoid air entrapment. The curled dough piece lastly passes under a pressure board to remove any gas pockets within and to cover the same.

Panning

The moulded dough pieces are immediately placed in the baking pans. Panning should be carried out so that the seam of the dough is placed on the bottom of the pan. This will prevent subsequent opening of the seam during proofing and baking. Optimum pan temperature is 90°F.

Final proving or proofing process

Final proving or proofing refers to the dough resting period during fermentation after moulding and the moulded dough pieces are placed in bread pans or tins. During this resting period, the fermentation of dough continues. The dough finally proofed or fermented in baking pan for desired dough height. It is generally carried out at 30-35ºC and at 85% relative humidity. Proofing takes about 55-65 minutes. During proofing, the dough increases extraordinarily in volume. The dough expands by a factor of 3-4 during proofing. During proofing, care has to be taken that the skin of dough remains wet and flexible so that it does not tear as it expands. A high humid condition is also required to minimize weight loss during proving. Temperature, humidity and time influences proofing. Proof temperature depends on the variety of factors such as flour strength, dough formulation with respect to oxidants, dough conditioners, type of shortening, degree of fermentation and type of product desired. During proofing, lower humidity gives rise to dry crust in the dough. Excessive humidity leads to condensation of moisture.

The baking process

After proofing, the dough is subjected to heat in a baking oven. Baking temperature generally varies depending up on oven and product type but it is generally kept in the range of 220-250ºC. During baking, the temperature of dough center reaches to about 95ºC in order to make sure that the product structure is fully set. When the dough is located in the oven, heat is transferred through dough by several mechanisms such as convection, radiation, conduction and condensation of steam and evaporation of water. Heat transfer inside dough is said to occur through the mechanism of heat conduction and evaporation/ condensation. The baking time of bread may range from 25 to 30 minutes depending up on size of bread loaf. After baking, bread is cooled prior to packaging to facilitate slicing and to avoid condensation of moisture in the wrapper. Desirable temperature of bread during slicing is 95-105°F.

Formation, expansion and trapping of gases

The gases primarily responsible for leavening baked goods are carbon dioxide, which is released by the action of yeast and by baking powder and baking soda; air, which is incorporated into doughs and batters during mixing; and steam, which is formed during baking.

Baking sets the final structure to baked goods. It involves simultaneous heat and mass transfer phenomena. The heat travels from the surrounding air into the interior of the dough or batter while moisture and other liquid compounds travel/escape from the main towards the exterior or surrounding air due to evaporation.

During mixing and in contact with liquid, these two form into a stretchable substance called gluten. The coagulation of gluten is what happens when bread bakes; that is, it is the firming or hardening of these gluten proteins, usually caused by heat, which solidify to form a firm structure.

Bakers turn to whey proteins because of the flavor, function, and nutritional aspects. Quick breads, cakes, muffins, bars, cookies, and other delicate baked goods may contain whey protein because it provides a neutral flavoring, not adding any undesirable tastes to the baked item.

Starch gelatinization is the process where starch and water are subjected to heat causing the starch granules to swell. As a result, the water is gradually absorbed in an irreversible manner. When it is cooked in boiling water, the size increases because it absorbs water and it gets a soft texture.

Water evaporates at 100°C, so once the bakery product dough has reached that temperature, moisture will evaporate and leave the bakery product. Moisture might not be able to leave easily though if it is trapped inside the dough. Longer baking times, means more evaporation of water, means crispier product.

Shortening melts at a higher temperature, so unlike butter, the formed dough will stay high longer and not melt so quickly with a melting point of 117° F, it is almost always in its solid, fluffy form. Bakery products made with butter, when not chilled before baking, will result in flat bakery products with a bigger spread.

Brown sugar can make bakery products soft and chewy, whereas white sugar might make them hard and dry. Do use brown sugar with baking soda in cookie recipes. Brown sugar is somewhat acidic, which means that it reacts with baking soda to produce carbon dioxide.

In text Questions

1. During mixing and in contact with liquid, a stretchable substance formed is called _____.

2. _____ should be carried out so that the seam of the dough is placed on the bottom of the pan.

3. _____ is the process where starch and water are subjected to heat causing the starch granules to swell.

4. Water evaporates at _____°C.

5. What happens during the baking process?

6. What traps air or gas in the dough during the baking process?

7. What does protein do in baking?

8. What happens to starch during Gelatinization?

9. How does water evaporate during baking and at what temperature?

10. What is a shortening in baking? What is the melting point of shortening?

11. Explain the basic concepts of baking process in detail.

CHAPTER 14

CAKES AND ICING

Meaning

A sweet food made with a mixture of flour, eggs, fat, and sugar.

Types of Cakes

There are few Examples

- Butter Cake
- Pound Cake
- Sponge Cake
- Genoise Cake
- Biscuit Cake
- Angel Food Cake
- Chiffon Cake
- Baked Flourless Cake

Cake Making Methods

There are five main methods of cake making:

- Rubbing–in
- Melting
- Creaming

- Whisking
- 'All in one' - same proportions as the creaming method

Each method produces products that have a different texture. The different proportion of ingredients used will determine the texture and taste of the cake product. The amount of fat in the cake product will determine how long the cake will stay fresh – without drying out. Cakes cannot be made successfully with low fat spreads; this is because they contain a higher proportion of water.

1. Rubbing in Method

- Air is trapped in sieving the flour and by lightly (with finger tips) rubbing the fat into the flour.
- Raising agents in the flour help the cake to rise. Examples of cakes made using the rubbing in method.

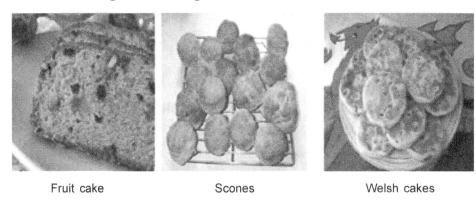

Fruit cake   Scones   Welsh cakes

Fig. 16. Examples of cakes made using the rubbing in method

2. Melting

- Fat and sugar ingredients are melted in a saucepan
- Texture tends to be much heavier than other cakes and will not rise much
- Bicarbonate of soda can be used to create a lighter texture.

Fig. 17. Examples of cakes made with melting method

3. Creaming

- Air is trapped by creaming the sugar and fat together
- This gives a lighter texture

4. Whisking Method

- Eggs and sugar whisked together to trap air (aerate).

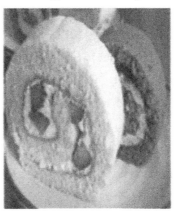

Chocolate vanilla swiss roll   Sponge flan

Fig. 18. Examples of cakes made using the whisking method

5. All in one method

If a baker wants to save on washing up, go for a simple all in one bake. Exactly as the name suggests, all the measured ingredients go into the bowl together and the mixing is done in a matter of minutes.

RICH CAKE

Ingredients

Margarine -200 g

Butter -200 g

Egg – 9 nos.

Cake flour -400 g

Black jack -90 g

Dried soaked fruits -900 g

Mixed spices -16 g

Castor sugar - 400 g

Baking powder -9 g

Fig. 19. Rich cake

Method

- Cream the fat on the working table and add in the sugar, continue creaming.
- Slowly start adding the eggs a little at a time, creaming it well.
- Once all the eggs have been used in the creaming process add in the black jack and mix it well.
- Add in the soaked fruits to the above mixture and fold in the flour mixed with the baking powder. Fold the mixture gently.
- Pour them into lined cake molds and bake them at 175°C till done.
- Cool the cake and serve it as required.

Fairy cakes

Victoria sandwich

Fig. 20. Examples of rich cake

PLUM CAKE

Ingredients

Plums Without Stone – 1 g

Grounded Almonds – 150 g

Bread Crumbs – 200 g

Sugar – 150 g

Cinnamon – 1 Teaspoon

Butter – 50 g

Flour – 300 g

Butter – 200 g

Sugar – 100 g

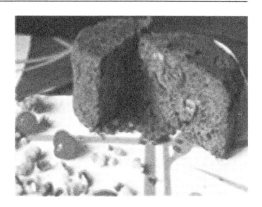

Fig. 21. Plum cake

Method

Prepare the base from

Flour – 300 g

Butter – 200 g

Sugar – 100 g

Knead the ingredients together and roll the dough to the size of the baking tin about 4 mm thick.

Prepare the covering by mixing

Ground almonds-150 g

Bread crumbs-200 g

Sugar-150 g

Cinnamon-1 teaspoon

• Add 2 tablespoons of apple juice and produce crumbles to cover the plums.

- Cut the plums in halves and put them like tiles on the base, sprinkle with ground cinnamon and sugar.

- Cover the plums shown and add small pieces of butter on top.

- Bake for 60 minutes at 160°C let the cake cool in the tin serve with whipped cream.

PINEAPPLE UPSIDE DOWN CAKE

Ingredients

For the topping:

Softened butter-50 g

Light, soft brown sugar -50 g

Pineapples -7 rings in syrup, drained and syrup

Glace cherry

For the cake:

Softened butter -100 g

Golden castor sugar- 100 g

Self-rising flour- 100 g

Baking powder -1 tsp

Vanilla extract-1 tsp

Eggs - 2

Method

- Heat oven to $180^0C/160^0C$

- For the topping, beat the butter and sugar together until creamy.

- Spread over the base and a quarter of the way up the sides of a 20-21 cm round cake tin. Arrange pineapple rings on top, then place cherries in the centers of the rings.

- Place the cake ingredients in a bowl along with 2 tbsp of the pineapple syrup and, using an electric whisk, beat to a soft consistency.

- Spoon into the tin on top of the pineapple and smooth it out so it is level.

- Bake for 35 mins.

- Leave to stand for 5 mins, then turn out onto a plate.

- Serve warm with a scoop of ice cream.

ICING

Powdered sugar, also called confectioners' sugar or icing sugar, is a finely ground sugar made by milling granulated sugar into a powdered state. It usually contains a small amount of anti-caking agent to prevent clumping and improve flow. Always sieve the icing sugar before use to avoid lumps.

Icing sugar, as the name suggests, is used in icings, confections, drinks etc. Its fine texture makes icing sugar ideal for dusting over cakes, pies and pastries to sweeten as well as to add an attractive decorative touch.

In text Questions

1. _____ can be used to produce a lighter texture.

2. Baking of the rich cake should be done at _____ .

3. The raising agent in rich cake is _____ .

4. Mention the different types of cakes.

5. What are the main methods of cake making?

6. List out the ingredients required for making rich cake.

7. Write the method of making the rich cake.

8. Write the difference between plum and rich cake.

9. List out the ingredients required for making plum cake.

10. Write the method of preparation of Pineapple upside-down cake.

11. How icing sugar is prepared?

12. What is the use of icing sugar?

CHAPTER 15

COMMERCIAL BREAD MAKING METHODS

The knowledge of Bread making dates back to several thousands of years. Over the years, Bread making method has evolved, to meet the production and quality requirements from time to time. While the conventional methods of bread making are still popular, technological advancements during the last about fifty years have helped to develop new methods which are gaining wide acceptability in different parts of the world and also in our country.

Process Steps

The production of bread consists of a number of different processing steps. These are as follows, in the sequence in which they are done.

Sifting of flour ⟶ Preparation of Dough Mixing/Mixing & Fermentation ⟶ Dividing

Moulding ⟵ Intermediate Rounding & Proofing ⟵ Fermentation

Panning in oiled Baking Tins ⟶ Final Proofing ⟶ Baking

Slicing & Packing ⟵ Cooling ⟵ Depanning

Fig. 22. Bread Making Process

1. Sifting

This is the initial step where sifting of flour is done to remove any foreign materials that might be present in the flour. The sifting step also helps to aerate the flour.

2. Preparation of Dough

The next step is the preparation of the dough. There are different methods of preparation of dough, which involves either Mixing alone or Mixing and Fermentation. The dough preparation stage is a very important step in bread making and the different methods of dough preparation are generally referred to as different methods of Bread Making.

3. Dividing

The prepared dough is then taken for dividing into pieces of required size. Dividing is done either manually or using dividing machine.

4. Rounding and Intermediate Proof

After dividing, the dough pieces are rounded, to ball shape. The rounded dough pieces are passed in to the interproofer. Here the dough pieces get time to recover their extensibility so that they can be moulded without breaking the surface skin to avoid stickiness and to get proper moulding. These two stages are usually bypassed when dough is manually divided.

5. Moulding

Here, first of all the dough is sheeted by passing through a set of pairs of rollers. Sheeting is done for expulsion of the trapped gases produced during the previous stages of processing. The sheeted dough is then passed through a pressure board during which they get moulded into a cylindrical shape.

6. Panning

At this stage, the moulded dough pieces are placed into the bread baking tins, which are greased properly with refined oil or emulsion. Panning is done in such a way that the dough pieces are centrally placed in the tins with the seals facing the bottom. This will prevent subsequent opening of the sealed dough during final proofing.

7. Final Proofing

The panned dough pieces in baking tins are then transferred into Final Proofer. The final proofer is a closed chamber, where the required heat and humidity are provided for fast rising of the dough in the tins without surface drying. Final proofing takes normally about 50-60 mins. During the final proofing, the panned dough pieces gradually rises inside the baking tins due to a gas produced by yeast action and by the end of this stage, the dough pieces rises to the required level.

8. Baking

The fully proofed dough pieces, then transfer to baking oven. When the dough pieces enter the oven after final proofing, the activity of yeast still remain for a short period and at a faster rate due to increasing temperature. The dough pieces increase in volume rapidly because of an increase in the rate of release of CO_2 gas and gas expansion. This rapid increase in the volume is known as "Oven spring". As baking proceeds yeast gets deactivated, the proteins get coagulated, starch gelatinizes and set the structure of the product after which there will be no further increase in volume. Towards the end of baking, golden brown color develop on the top and sides of the bread.

High temperature short time baking is considered ideal for getting soft bread. The normal baking temperature ranges from 220-260°C, depending on the size and variety of bread.

9. De-Panning and Bread Cooling

Once the breads come out of the oven, they are de-panned and are stacked on trolleys. For the purpose of slicing and wrapping, loaves must be cooled. Faster cooling can be achieved by forced circulation of air over the loaves. At the end of the cooling process, the temperature of the interior of the loaf shall be nearer to the room temperature. The humidity of the air in the cooling may be controlled to prevent excessive moisture loss from the surface.

10. Slicing and Packing

The sufficiently cooled breads are sliced and packed in polypropylene pouches.

Different Methods of Bread Making

As mentioned before, there are different methods of Bread Making (ie. methods of dough preparation). The bulk fermentation method is the traditional one, followed and evolved over thousands of years of the history of bread making. The other two methods are the result of technological advancements during the last 50 years.

Bread Making Methods

These can be broadly grouped into three based on the way the dough is prepared.

I.   Conventional Methods

II. Chemical Dough Development (CDD) Method - Bulk fermentation method

III. Mechanical Dough Development (MDD) Methods

   (a) Straight Dough method

   (b) Sponge & Dough method

(c) Emergency Dough method

(d) Delayed Salt method

I. CONVENTIONAL METHODS

The Conventional Methods are characterized by long fermentation periods. Depending upon the way, the mixing is done and the ingredients are added, these methods are further classified as discussed below:

a) Straight Dough Method

(i) Mixing

This is the basic conventional method of bread making from which all the other methods are obtained. This method consists of a single mixing process after adding to flour required quantity of water and all other ingredients such as yeast, salt, Vanaspati, preservatives, emulsifiers, improvers, etc. The mixing is continued till the dough is developed to a smooth consistency. The dough finish is kept somewhat firm to allow for softening of dough during the subsequent bulk fermentation period. Conventional slow speed double arm or single arm mixers or planetary mixers are generally used. The Mixing time ranges from 15 - 30 minutes, depending upon the type and speed of mixer and batch size. Mixing is to be continued till elastic dough with a smooth, silky and non-sticky finish is obtained. Spiral mixers or Horizontal mixers can also be used for faster dough development. The desired dough temperature at the end of mixing is about 80° F. The temperature of water needs to be adjusted to get the desired dough temperature. During mixing the flour, proteins get hydrated and form a combined elastic protein called gluten.

(ii) Bulk Fermentation

The mixed dough is then kept for Bulk Fermentation, generally for about $1^{1/2}$ to $2^{1/2}$ hours. The duration of bulk fermentation depends upon the quality of flour used, with stronger flour requiring longer fermentation. Other major factors determining the time of bulk fermentation are the quantity of yeast used and dough temperature. The mold and rope inhibitors added to the dough as well as added sugar and salt depresses the fermentation rate. During fermentation, the pH drops from 5.3 to as low as 4.5 due to the formation of organic acids. The lowering of pH has a marked effect upon the hydration and swelling of the gluten, rate of enzyme action and other chemical reactions involving organic compounds.

The lowering of pH and action of proteolytic enzymes on protein, production of organic alcohols etc., alters the colloidal structure of proteins in dough. This helps to increase the extensibility and elasticity of the gluten, thereby enabling it to form thin gas retaining cells which can expand without rupturing. A dough with maximum gas retaining capacity and has developed maximum elasticity and springiness is said to be mature. Lower temperature prolongs the fermentation time and higher temperature may lead to wild fermentation. The temperature of 80 – 85° F and RH of 75% is optimum.

It is a normal good practice to activate yeast before adding to the flour. Activation of yeast is done by dispersing yeast required to be added in each batch, in a bucket or suitable vessel, with enough water (about 1 lit water per kg yeast). Some quantity of sugar and flour are also added. The dispersed yeast is kept for about 20 minutes to activate. This helps to reduce the quantity of yeast and also to get faster fermentation. The fermenting dough is generally given one or two knock backs which improve the gas retaining property as well as the rate of gas production. Knockback refers to punching of the gas built up during the fermentation period. It is generally given after the lapse of about 60 to 65% of the fermentation period.

Advantages of Straight Dough Method

It is a simple process. It requires lesser processing time than sponge dough method. The power and equipment requirements are larger from investment point of view. This method gives characteristic bread flavor to the product.

b) Sponge and Dough Method

Another conventional method, the popular method is the Sponge and Dough Method. This method consists of two steps mixing. At the first mixing stage, some portion of the flour (about 65 - 70% of total), yeast (full or part), small quantity of sugar (about 1- 2%) and required quantity of water are mixed together. The mixed mass is called' Sponge'. The consistency of sponge can vary from soft to stiff depending upon the baker's preference. Mixing is done just to blend all the ingredients to a smooth homogeneous mass. As mixing is not done to full extent of smooth finish, at this stage, only part development of gluten is achieved. This will enable the sponge to retain some gas, and rise in volume during fermentation. The temperature of sponge is desired to be at about 80° F for optimum results.

The sponge is allowed to ferment for 12 to 16 hours depending on the flour quality and quantity of yeast. The fermentation takes place faster than that in straight dough method as yeast inhibiting agents such as salt are absent.

At the end of the fermentation, the sponge is again mixed with the remaining ingredients, namely balance quantity of flour, sugar, vanaspati, emulsifiers, water, etc. This second mixing is done until the dough is fully developed to a uniform, smooth, shiny and non-sticky finish. The developed dough is then kept for a short second fermentation called "Floor Time". Floor time ranges from 15-20 minutes and this improves the machinability of the dough. Thereafter the dough is taken for further processing.

Advantages and Disadvantages

Sponge and Dough method helps to get bread with better volume and internal (crumb) characteristics. It is more flexible with respect to mixing and fermentation and also helps to save on yeast. Can handle weaker flour for bread production. But the process requires more plant space. Also power consumption and machinery requirements are more. The fermentation losses are relatively more.

c) Delayed Salt Method

This method is a modification of Sponge and Dough method. Here the entire quantity of flour for a batch is taken at the first mixing stage itself, thus it is a 100% Sponge method. After first mixing, bulk fermentation is given for 2 - 3 hours depending upon flour quality. After fermentation, the dough gives a second mixing during which all the salt is added towards the end of the mixing period. No flour is added in the second mixing. The mixed dough gives time and processed further. This method can be used for processing strong flour, which otherwise would require much longer mixing and fermentation time.

d) No Time Dough Emergency Dough

This method is used only in the case of an emergency situation when dough has to be made ready for processing in a very short period of time. Thus, it is used typically in situations such as machine breakdown or last minute urgent orders. The method typically involves straight dough mixing. The dough is made using a much higher quantity of yeast - almost double the quantity. The dough temperature is also kept relatively higher at about 28 - 30°C. After mixing, the dough is taken immediately for further processing.

Advantages and Disadvantages of Emergency Dough

The advantage of this process is that bread can be made in a relatively short period of time. However, the quality of bread is usually inferior to that produced from the normal conventional process with respect to internal characteristics. It has higher ingredients, cost and does not have a good bread aroma.

II. CHEMICAL DOUGH DEVELOPMENT (CDD) METHOD

In this method, the required development of dough is achieved by use of certain chemicals, typically reducing agents. The dough is mixed in normal mixer and all the ingredients are added together. After mixing the dough is taken for subsequent processing without any fermentation. The development of dough is achieved by addition of reducing agents such as L-cysteine hydrochloride and sodium metabisulphite which assist in softening of gluten. These chemicals also reduce the time required for mixing of the dough. These reducing agents are used in low level and are often used in conjunction with usual bread improvers having components like Ascorbic Acid and Potassium Bromate. The advantage of the Chemical Dough Development method is that it does not require special-costly machines for fast mixing. It also does not require high energy consumption. The process is faster than conventional method and produces reasonably good quality bread. However, the bread lacks in flavor and aroma as compared to conventionally processed bread.

III. MECHANICAL DOUGH DEVELOPMENT (MDD) METHOD

According to this method, the normal development of dough, achieved during long mixing and the bulk fermentation period in the conventional methods, is achieved by intense mixing and high energy input. The most common one is the Chorleywood Bread Process (CBP) originated in the UK. This method uses specially designed high speed mixers, which are capable of completing the dough development in a much shorter mixing time of 3 to 5 minutes. The mixing intensity is such as to impart an energy input of about 11 kWh kg of dough in less than

5 minutes. The method involves adding all the flour, water and other ingredients into the mixer bowls and straight mixing for about 3 minutes. The mixed dough is well developed and is immediately taken for further processing of dividing and baking. This process has become popular in our country also and most of the bakeries in the northern part of the country follow this method.

Advantages and Disadvantages

This method enables reduction in the total production time. Generally, the internal characteristics of bread produced are much finer. Because of more addition of water, higher yield of bread is achievable. As the mixing time is within a short range the dependence of bread quality on the skill of mixing operator is less. The requirement of space is much less compared to conventional method. The process requires special mixer with High energy input and power consumption. Yeast consumption is higher. The requirement of chilled water, with low water temperature is essential. The bread produced does not have characteristics, bread aroma, as compared to conventional method.

Continuous bread making methods

This method comes under the broad category of MDD. The process involves continuous dough preparation of MDD under pressure with a continuous flow of ingredients. The developed dough is extruded into baking tin directly instead of moulding. Normally, a preferment brew is used in these processes to get the desired flavor. There are two common methods of producing bread continuously.

Do-Maker Process

In this process, a brew is made by stirring a mixture of water (60%), sugar (8.0), yeast (3.0), yeast food (0.50), salt (2.0), and mold inhibitor. The brew is fermented for 3 to 4 hours during which temperature rises by 10 - 15 ° F and pH drops from 6.5 to 4.7. The fermented brew, other ingredients, flour and remaining water are metered into a mixer. The dough is mixed and transferred to a dough developer through dough pump. Once the dough is developed with intense energy input, it is pumped to an extruder where it is extruded into the greased baking tin. It is then proofed and baked as per the conventional procedure.

Amflow Process

The Amflow method uses a flour based brew/liquid sponge instead of non-flour brew used in the Do Maker process. Flour brew consists of some flour, water, yeast, salt and sugar. The use of part of the flour in the brew help in cost saving in ingredients since flour is fermented rather than sugar. This also helps to improve the flavor. After fermentation for about 2 - 3 hours, the ferment is pumped into the pre-mixer along with melted fat, oxidants, sugar solution, remaining flour, water and other ingredients. The mixed dough is pumped into the developer through a dough pump. The dough is developed by incorporating the required energy. Continuous bread making system reduces processing time, floor space and some labour. The bread made by these methods has fine uniform grain structure. These methods are used for very large scale production involving 2000 - 3000 kg dough per hour.

EVALUATION OF BREAD
Bread Faults

Bread faults, which can be many in number, may be broadly divided into two main groups, namely External Faults and Internal Faults. These faults may originate from poor quality of ingredients or from a faulty production process, or a combination of both. The occurrence of bread faults will not be frequent and their effects not severe. If proper precautions are taken to ensure, use of proper quality ingredients, use of right formulations, adherence to proper process conditions at different stages of production and that equipment are kept at optimum operating efficiency. It is convenient to list them according to bread quality characteristics and indicate various possible causes and remedial actions for the benefit of bakers.

External Faults

Nature of fault sand some probable causes to be corrected

Lack of Volume

a) Insufficient Yeast

b) Insufficient water absorption

c) Too much salt

d) Excess of diastatic activity in flour

e) Over-mixing or under-mixing of dough

f) Over- fermented for under- fermented doughs

g) Too Iow a dough temperature

h) Insufficient final proof

i) Improper humidity conditions during proofing

j) Insufficient dough weight for baking tin size

k) Excessive oven temperature

l) Weak flour

Excessive Loaf Volume

a) Insufficient salt

b) Over-aging of dough

c) Over proofing

d) Too much dough for pan size

e) Low oven temperature

Pale Crust Color

a) Insufficient sugar

b) Deficiency of diastatic activity

c) Too high a fermentation temperature

d) Skinning of dough during proofing

e) Dry proof box

f) Over- fermentation of dough

g) Low oven temperature

h) Low top heat in oven

i) Too short a baking period

Crust Color Too Dark

a) Excessive sugar in formula

b) Immature dough

c) Too high an oven temperature

d) Excessive top heat in oven

e) Over-baking

f) Oven atmosphere too dry

Crust Blisters

a) Improper mixing

b) Immature dough

c) Incorrect moulding

d) Excessive steam in proof box

e) Excessive oven steam leading to condensation

Excessive Crust Thickness

a) Insufficient sugar

b) Deficient diastatic action

c) Skinning of dough during proofing

d) Over-aged dough

e) Low oven temperature

f) Too long a baking period

Shell Tops

a) Immature flour freshly milled

b) Deficiency in diastatic action

c) Excessively stiff dough

d) Immature dough

e) Insufficient Final proofing

f) Dry oven steam

g) Skinning of dough during proofing

Uneven Shape

a) Flour with high maltose figure

b) Excessive amylase addition

c) Improper moulding

d) Improper panning placement in baking tin

e) Improper final proofing

f) Improper depanning and handling of hot bread

g) Use of blunt slicing blades

h) Damages during slicing and packing

Internal Faults

Nature of fault sand some probable causes to be corrected

Dull Crumb Color, lacking brightness

a) Improper dough mixing

b) Too long final proof

c) Over fermentation of doughs

d) Improper dough development

e) High fermentation temperatures

f) Excessive use of dusting flour

g) Excessive use of oil at divider

h) Improper moulder setting

Coarse Grain (Non-uniform cellular structure)

a) Weak Flour

b) Very stiff doughs

c) Too high water absorption (slack doughs)

d) Over - mixing

e) Under fermentation

f) Improper moulding

g) Insufficient dough weight for pan size

Poor Texture (Thick cell walls)

a) Very stiff doughs

b) Improper mixing

c) Excessive amylase action

d) Over fermented doughs

e) Skinning of sponge or dough during fermentation

f)   Excessively high proof box temperature

g)   Over-proofing

h)   Insufficient dough weight for pan size

i)   Incorrect use of bread improver

Poor Flavor

a)   Low quality ingredients (flour, yeast, etc)

b)   Insufficient salt

c)   Unbalanced formula

d)   Over fermentation / Under fermentation

e)   Excessive use of yeast

f)   Over proofing

g)   Under baking

h)   Poor bakery hygiene

i)   Use of old trough and pan greasing material

j)   Presence of external odors

Poor Keeping Qualities

a)   Unbalanced formula

b)   Poor quality ingredients

c)   Improper mixing

d)   Over fermentation / Under fermentation

e)   High dough temperature

f)   Oven final proofing

g)   Low oven temperature

h)   Improper bread cooling conditions

Holes in Bread

a) Use of freshly milled flour

b) Weak flour

c) Insufficient salt

d) Improper mixing

e) Excessively stiff doughs

f) Over fermented doughs

g) Skinning of sponge or dough

h) Improper molding

i) Excess divider oil

j) Too high a proofing temperature

Bread Staling

Bread, which has a moist and spongy crumb, is subject to a continuing deterioration in quality, particularly with respect to the firmness and eating quality, and this change is commonly called as 'Staling'. The staling occurs to both the crust and crumb portions of bread during storage. Actually, the staling process begins from the time the bread is cooled and continues progressively thereafter.

The crust, which is dry and crisp when fresh becomes softer and leathery upon staling. It also loses its appealing aroma. Whereas in the case of crumb staling, the texture becomes a film. The crumb also becomes harsher and crumbly as bread stales. The aroma and eating quality deteriorate on staling. As the staling progresses, the crumb also loses moisture. However, the qualities lost because of staling can be regained to a great extent by heating to a temperature above 60°C.

Retarding of Staling

Though there is no practical way for entirely stopping staling of bread. Various methods have been developed for retarding its rate. This is achieved by use of proper ingredients and use of various special additives. Bread made from weak flour, which have lower gluten content, stale much faster than that from stronger flour. Addition of vital wheat gluten markedly reduces the rate of staling. The addition of other ingredients such as milk solids, sugar, vanaspati and soya flour has also beneficial effect on staling rate. Flour with low amylase activity, gives bread that stales faster. Addition of fungal or bacterial alpha-amylase help to reduce the rate of staling on storage. Use of dough conditioners such as Glycerol Mono Stearate (GMS), Sodium Stearoyl Lactylate (SSL) and Datem Esters is known to bring about remarkable improvement in the retardation of bread staling and hence are widely being used in Bread industry.

Extrusion Dough Sheeting

A dough sheeter is a kitchen machine that rolls out pieces of dough to a desired thickness. The resulting sheets are smooth, uniform and completed in a few minutes, a much shorter turnaround than rolling by hand.

A countertop or a table top dough sheeter is a piece of industrial equipment that bakers can use to make the dough in large quantities without taking a lot of time. It is perfect for restaurants and bakeries with tons of orders on favorite foods, including pastries, pasta, and pizza.

A dough sheeter is an appliance used in food preparation, which flattens the dough into sheets. The general principle is that baker puts an oval sized ball into the top of the machine and a very uniformly rolled sheet will come out at the bottom. Dough Sheeters have used in a number of restaurants.

The dough is compressed between two or more rotating rollers. When done the right way, a smooth and consistent dough sheet is produced. The dough then passes one or several gauging rollers (mostly on conveyors) that reduce the dough to the required thickness. After this, the dough sheet is shaped into a desired dough product. This technology is mainly used in industrial production machines for (semi) industrial bakeries and the food industry. Most dough sheeters can handle a wide variety of dough, depending on the machine manufacturer. Most commonly dough sheeting technology is used for the production of laminated dough products like croissants and pastries, but it is also suitable for the production of bread, flatbread and pizza.

Function

- Shape the dough from individual dough batch to continuous dough sheet

- Less damaging of the gluten network

- Laminate layers of dough together (no pocket proofers and dividers are necessary as the dough sheet is the base of every product).

Benefits

A big benefit for using sheeting technology is the large dough capacity that can be handled. Dough sheeting manufactures are able to process high quality dough sheets at high capacities. Another benefit is that sheeting makes it possible to handle a great variety of dough types which traditional dough production systems can not handle, for example strongly hydrated wet and sticky dough.

Most dough sheeters can handle a wide variety of dough, depending on the machine manufacturer. Most commonly dough sheeting technology is used for the production of laminated dough products like croissants and pastries, but it is also suitable for the production of bread, flatbread and pizza.

Fig. 23. High Quality Dough Sheets

Fig. 24. Dough Extrusion Sheeter

Extrusion sheeting uses a twin-screw extruder with an innovative wide-slot die to produce a thin sheet of dough directly feeding a rotary cutter. The products are the same as those produced on a conventional dough sheeting line, but the equipment is considerably simpler and more flexible. Extrusion sheeting instead of conventional mixing, sheeting and gauging reduces both the capital investment needed for a snack cracker line and the floor space required.

The lines are simple to operate and, having fewer units, give significant reductions in cleaning time and maintenance costs. The system is suitable for any type of baked or fried snack that is cut from a sheet of dough. Wheat and maize are the most common ingredients, but many types of flour can be processed, either on their own or as part of a blend.

The process supports dedicated lines running at high output, and the low cost and flexibility of the line also makes the production of small batches of snack crackers economically viable. The extruder can be quickly and easily switched to make other snacks such as direct expanded curls and balls, or co-extruded filled pillows, bars and wafers.

Missing Questions

1. The four essential ingredients required for bread making are _____, _____ _____ and _____.

2. A living organism used in bread making process is _____.

3. Gluten is formed from _____ present in flour.

4. The process of _____ helps in gluten development.

5. The liquid used for bread making can be or _____ .

6. Fat acts as a _____ in bread.

7. Fermentation is a process in which acts upon _____ to form, _____ , water and _____ .

8. Why do we knead the dough? Explain in brief

9. What is fermentation? Explain in brief

10. Why is knocking back important? Explain briefly

11. Why are glazes applied on bread products?

12. Explain the commonly used glazes.

13. Explain the theory related to bread making process.

14. List the various raw materials used.

15. Explain their suitability and the purpose they perform in bread making.

16. Explain the process steps of bread making.

17. Describe the different methods of bread making and their salient features.

18. What do baker understand the reasons for common bread faults and suggest remedies?

19. Explain what do baker know about bread staling and how to retard the same.

20. What are the main constraints of baking industry?

21. Discuss the present status of milling & baking industry in India.

22. Classify bread and discuss in brief the role of ingredients in bread production.

23. Give a brief description of general bread making procedure.

24. How mixing is important in bread quality?

25. Discuss different types of mixing machine used in bread industry.

26. Why bread dough is subjected to molding, dividing and proving operations?

27. Discuss physical and biochemical changes in dough during mixing, fermentation and baking.

28. What is a dough sheeter?

29. What is a dough roller used for?

30. How does dough sheeter work?

31. Describe the different processing steps in bread making in a bakery unit.

32. Which are the different methods of bread making?

33. What is the difference between sponge and dough and delayed salt method?

34. What do baker understand by the term CDD?

35. What are the salient features of MDD?

36. Which are continuous bread making methods?

37. What are the major bread faults?

38. What is bread staling?

CHAPTER 16

BREAD ROLLS

INTRODUCTION

Think of eating and the baker think of bread. It can be in the form of a loaf baked in a bread form or a family molded bread baked open on a baking tray. Bread includes rolls, buns and bread sticks. The ready and mouthwatering aroma of a well baked bread is a very appetizing sight. No continental meal is really complete without bread as an accompaniment. Bread, as we already know, is a dough of wheat flour and water, seasoned with a little salt, raised by the action of yeast and then baked in an oven. But there can be many variations to this basic bread. Different types of flours, e.g. whole meal flour, barley flour etc. can be used and can also make spiced or sweetened, flavored or enriched breads.

Basics of bread making process

Before the baker learn to make bread, the baker must know the actual activity that happens in the dough. The four essential ingredients in bread are: flour, yeast, water, salt, sugar, fat and eggs are often added for nutritional value, better flavor, color and texture.

Breads are made by a fermentation process in which yeast, an one cell plant, feeds on sugar and converts it into carbon dioxide gas, water and alcohol.

Yeast and Sugar (from flour) →7 Carbon-di-oxide (gas) + Water + alcohol

The sugar required for the action of yeast comes from flour itself, which contains 1% sugar, and any sugar added during preparation. The fermentation process requires sugar and proper conditions of temperature and humidity. It then results in gradual expansion of dough and finally it is doubled in volume. In addition to yeast multiplication and activity, the gluten of flour must be developed. It is the gluten, which gives the dough elasticity or stretchability which is necessary for it is rising in volume. Gluten is formed from the proteins present in flour, on addition of water and salt. The process of kneading is accomplished by considerable manipulation by hand or by machine. As bread rises during the fermentation process, the gluten stretches to form the cellular structure of the dough which should be light and porous.

Raw materials used in bread making and their role

Although whole wheat flour contains protein, certain qualities of wheat have higher percentage of protein. This type of wheat flour is more suitable for bread making as it develops stronger gluten. Both fresh compressed yeast and dry yeast can be used for bread making, but fresh yeast gives better results. Dry yeast, if used, should be used in 50-70% quantity of the fresh one as it is more concentrated. Also activate the dry yeast before using.

To activate yeast: Add measured quantity of yeast to little lukewarm water and dissolve properly. From the weighed quantity of flour, add some to the dissolved yeast so that the baker gets a thin paste consistency. Keep this in a warm place for 10-15 minutes. If air bubbles start coming out, yeast is ready for use. If after half an hour also, no bubbles are visible the yeast is dead and inactive and should not be used. While sugar aids in rising of the dough, salt retards or controls gas formation. The dough would rise very quickly and the bread will not be of good quality. The liquid used for breads and rolls may be milk or water. The liquid dissolves salt, sugar and yeast and hydrates the flour. Milk is more often used in powder form as quality can be regulated better and storage is easier. Eggs add color and flavor, emulsify fats and produce a bread or roll of higher protein content. Solid shortenings (fats) in the form of butter, margarine or hydrogenated fats (vanaspathi ghee) act as tenderizers.

Other ingredients like cheese, onions, garlic, raisins, currants, candied fruits and peels are often added to produce different types of bakery products. All ingredients should be weighed in the correct proportions before starting. The bakery is more of a science as it is based on scientific principles, so exact proportions and right procedures must always be followed to produce good baked products. Right temperature for the dough is 200^0C or 80°F. Depending upon the atmospheric temperature the baker has to regulate the dough temperature by using cold water in very hot weather, tap water in normal conditions and lukewarm water in cold seasons. Bigger professional ovens have been proving chambers attached to them, where the conditions of temperature and humidity are controlled as per the requirement of yeast. The temperature in these chambers is controlled by production of steam which also keeps the crust from drying out. If provers are not available rolls may be proved successfully by placing the dough on racks, near an oven, but away from air drafts.

Different types of bread and bread products

There are many variations possible to a simple basic dough. Variations can be in the form of shapes or by addition or alteration of certain ingredients. By dividing dough into small and easily manipulated pieces, the baker can get a wide range of bread rolls or buns.

Similarly, the baker can take the bread in a closed tin, open tin or on a tray to change its shape. By adding certain sweet or savory ingredients, the baker can radically alter the bread's taste and texture.

128

The breads and rolls we are going to discuss in this chapter are:

1. Bread rolls

2. Hamburger rolls

3. Fruit buns

4. Basic Bread

5. Milk Bread

6. Brown Bread

i. Preparation of Bread rolls

These are generally eaten with continental meals to provide the-required cereal.

Ingredients

Flour	-	1 kg
Sugar	-	100 gms
Fat	-	100 gms
Yeast	-	20 gms
Salt	-	20 gms
Water	-	550 ml
Egg	-	1 no.
Calcium Propionate	-	3 gms

Procedure

• Take part of water and mix with yeast.

• Add little quantity of sugar and flour. Make thin paste and keep aside for about 10 min.

• Sieve the flour with powder form of the ingredient.

129

- Dissolve salt; left over sugar.

- Make a well in center and pour the yeast mixture and mix it well.

- Add salt; sugar; water gradually and prepare a smooth dough.

- Add egg and cream fat into the dough and well till gluten developed

- The dough will be mixed well.

- Ferment the dough for 30 min.

- Knock back the dough. Allow 30 min for relaxing.

- Mould into various fancy shapes and place them on a greased baking tray.

- When half proofed give egg wash, sprinkle some sesame seed, poppy seed on top.

- Bake at 220° c till golden brown color. After baking give oil wash for shining.

ii. Preparation of Bread rolls

Ingredients

Flour	-	225 g
Milk powder	-	5 g
Salt	-	2.5 g
Sugar	-	10 g
Fresh yeast	-	10 g
Butter or Margarine	-	10 g
Egg	-	1 (for egg wash)

Method

1. Sieve flour and milk powder together onto a marble table top. Make a bay in the center of the sieved flour.

2. Dissolve yeast in a little (about 40 ml) water in a mug.

3. Dissolve salt and sugar separately in another mug using about 40 ml water.

4. **Mixing the dough:** Pour yeast solution at the center of the flour and mix with one hand. After the whole liquid has been mixed with flour, add sugar and salt solution mixing in the same manner.

5. **Adjusting the consistency:** Add more water if required so that a soft shaggy mass is formed. The dough at this stage should neither be soft nor too hard. If it feels very dry and stiff, mix in a little more water; if it is too loose and wet, work in some more flour.

6. **Kneading:** Once the consistency of the dough has been corrected, the baker has to knead it. The process of kneading has already been explained earlier.

7. **Addition of fat:** Cream butter or margarine on a tabletop, using the heel of the baker's hand. Mix the fat into the dough and knead again for about 5 minutes to mix fat properly.

8. **First fermentation:** Place the dough in a greased bowl and cover with a moist duster and keep the bowl in a warm place. Until the dough has doubled in bulk. The time required will vary from 45 minutes to $1^{1/2}$ hours. To test that the dough has risen enough, press a finger into it. If the depression remains, filling in only very slowly, the dough is ready.

9. **Knock-back:** Turn the risen dough out on the work surface and punch it to reduce the volume. Then knead it until firm. This distributes the air bubbles in the dough and gives better texture to the finished products.

10. **Second fermentation:** Keep the dough again in the bowl, covered with a moist duster till the volume of the dough is doubled again. This time it will take 35-50 minutes.

11. **Second knock-back:** Knock-back as explained at point No.9.

131

12. **Dividing the dough:** Shape the dough into a long cylinder and divide to get eight equal portions. The best method is to divide the cylinder into two and then further keep making halves till the baker get eight equal balls.

13. **Intermediate proving:** Shape each dough piece into a tall using a light, even pressure with the heel of the baker's hand and turning it in a clockwise direction, pleats, if any should be towards the bottom. Keep the balls covered with a moist duster for five minutes for intermediate proving.

14. **Readying the baking tray:** Take a clean baking tray and grease it lightly with oil. Remove any extra oil by rubbing with paper.

15. **Shaping the balls:** Take the ball the baker had made first and give it a desired shape. Keep on the greased baking tray. Shape all balls and keep in rows on the baking tray with a gap of 2" on all sides. The gap between various rolls is important as they will be increasing in size and will stick to each other if enough space is not provided. This will spoil the appearance.

16. **Final Proving:** Cover the trays with duster, keep in a warm place till the rolls are doubled in volume.

17. **Egg wash:** Beat one egg slightly and using a pastry brush, lightly coat the rolls.

18. **Bake:** Bake is an oven preheated to 200°C for about 20-25 minutes or till golden brown in color.

19. **Cool:** Keep on a cooling rack for cooling and brush a little butter on top. Bread rolls can be shaped variously and the baker can develop as many as the baker's imagination permits. A few shapes are discussed here.

   (i) **A simple round:** Bread rolls can be shaped as rounds, which are simple and quite acceptable. The baker can also make a crisscross pattern by drawing deep lines using a sharp knife before egg wash.

(ii) **Single knot:** Exerting an even pressure with both hands, roll each portion if dough backwards and forwards until it is about 9" in length. Tie the strip of dough into a loose single knot.

(iii) **Double knot:** Roll the dough with both hands till 12" in length. Bring both ends towards the center and the two adjacent single knots.

(iv) **Clover leaf rolls:** Divide one ball of dough into 3 equal parts. Make 3 rounds and place on greased tray to form a clover leaf pattern.

(v) **Coiling a turban:** Roll out a cylinder of 12" length and coil into a spiral.

(vi) **Plait:** Divide the dough into 3 equal parts and roll each one to 5-6" length. Using the three strands, tie a plait. At the beginning all 3 stands should be joined neatly and the same should be done at the end.

In text questions

1. The baking temperature for Bread roll sheet is_____.

2. After shaping the roll, it should be put in _____ for setting.

3. The Bread roll should be covered with _____ and _____ .

4. List out the ingredients required for making Bread roll. Explain their role.

5. How to judge the quality of rolls?

6. What are the different methods of preparing bread rolls?

7. Write the quantities of the ingredients used in making Bread roll.

CHAPTER 17

PIZZA

This is a tried and tested veg pizza recipe and if everything is followed correctly, the baker can able to make great pizzas with a choice of toppings. This homemade veg pizza beats all the pizzas in Indian metros like the Domino's pizza, smokin joes and even pizza hut.

This recipe of homemade veg pizza is a pizza recipe made from scratch. Except for the cheese, no ready-made sauces or pizza bread bases are used. This is a basic veg pizza. For the flavorings of the sauce, dried oregano and basil can be used. However, can use fresh oregano and basil too. But this is the best sauce. The pizza bread is made from maida (all-purpose flour). Wheat flour can be used too. But wheat flour will make the pizza bread base a little dense.

It is divided into 3 main parts:

1. Making the pizza bread
2. Making the pizza sauce
3. Assembling pizza

Step I – Making the pizza base

1. Dissolve the sugar in warm water and add dry active yeast to it.
2. Stir the whole mixture and keep it aside for 10 to 15 minutes.

3. When the yeast is getting doubled up, take a cup of flour with salt in a bowl. Mix well. Add olive oil.

4. After 10 to 15 minutes, the yeast will bubble and will see a good frothy layer on top.

5. Now add the frothy yeast mixture to the flour. With a whisk or wooden spoon, stir the mixture.

6. Add another cup of flour and keep on stirring.

7. The mixture will become sticky.

8. Add the remaining flour and continue to stir.

9. The dough will leave the sides of the bowl but still will be a bit sticky.

10. Knead the dough with the hand. Knead the dough to a smooth dough. Dust with flour to prevent sticking. The dough will be soft and elastic. Apply some olive oil to the dough all around.

11. Keep the dough in a deep large bowl and cover loosely with a kitchen napkin or towel for 1.5 to 2 hours.

12. After 2 hours, the dough has beautifully risen and doubled up to what we see below.

13. If a baker plan to make the pizza right away, then flatten the dough into a disk. Then start rolling the dough.

14. The dough gets leavened, even in the fridge.

 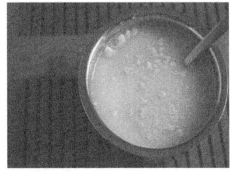

Fig. 25. Step I – Making the pizza base

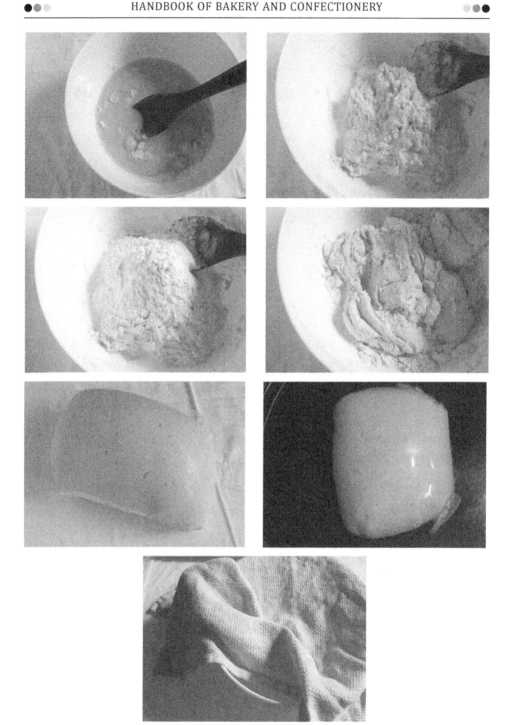

Fig. 25. Step I – Making the pizza base

Making pizza sauce

1. Chop the tomatoes directly or blanch them and then chop the tomatoes. If the baker chose the second method of blanching, as then less cooking is involved when making the sauce. Instead of chopping the baker can also puree the tomatoes.

2. In a pan heat olive oil and add chopped garlic to it. Fry the garlic for a minute.

3. Now add the chopped blanched tomatoes. Stir and let the tomatoes cook for 5 minutes more.

4. When the tomatoes are getting simmered, slice the onion, capsicum and olives.

5. Instead of sautéing, the veggies lightly marinated for 15 minutes with olive oil, basil and oregano.

6. The tomatoes are softened now. Add the herbs, salt, crushed black pepper.

7. Mix well and cook for a few more minutes.

8. The pizza sauce is ready.

Making veg pizza

1. Preheat the oven to 200°C. Grease the baking pan with olive oil and then dust it with maize flour (cornmeal), semolina (rava or sooji) or flour.

2. Take a large ball from the dough. Can use the whole dough to make an extra-large pizza or even make two large pizzas.

3. Flatten the dough into a disk. Now on a floured surface with a rolling pin, roll the dough ball into a large round or oval shape of about 1/4 inch thickness. Roll the dough from the center towards the outer edges. Dust the surface with flour. The dough is extremely elastic. So be careful while rolling the dough.

Fig. 26. Step II – Making pizza sauce

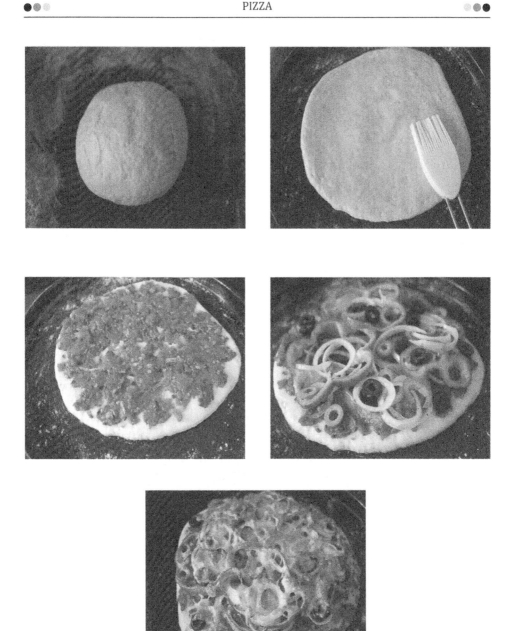

Fig. 27. Step II – Making veg pizza

4. Place the pizza base in the baking pan. Be careful as the dough has so much elasticity, it just stretches more when baker place it. Do this work by placing very gently. Brush the surface with olive oil.

5. Spread the tomato pizza sauce on the pizza.

6. Top up with the veggies and olives.

7. Top with some shredded mozzarella cheese or pizza cheese.

8. Bake the veg pizza in the oven for 10-15 minutes at 200°C till the base becomes golden brown and the cheese on top melts and gets browned.

9. The veg pizza is ready and out of the oven.

10. Serve the veg pizza immediately with tomato sauce or tomato ketchup and the herbs.

Making pizza dough

- Sprinkle sugar in warm water.

- Add yeast. Stir and let the mixture set at room temperature for 10-15 mins till it becomes frothy.

- In a bowl, add one cup flour, salt, olive oil and the frothy yeast mixture.

Fig. 28. Baked pizza

- Stir. Add another cup of flour. Stir again. The mixture becomes sticky.

- Add the last cup of flour and continue to stir.

- Knead the dough into a smooth, springy ball. Apply some olive oil all over the dough.

- Cover loosely and keep in a large bowl at room temperature for 1.5 to 2 hours.

- The dough will double up and nicely leaven.
- Making pizza sauce.
- Blanch the tomatoes and the chop them.
- Heat oil. Fry the chopped garlic. Add the tomatoes and wine. Let them cook uncovered for 4 to 5 minutes.
- Add the herbs, salt and pepper. Mix well. Cook further for 3 to 4 minutes.
- Assembling and making veg pizza.
- Flatten the dough to a disk. On a floured surface, roll the dough.
- Place the dough onto a greased and dusted pan.
- Brush some olive oil on the pizza base. Spread the tomato sauce on the pizza.
- Top with the veggies. Spread the grated cheese.
- Bake the veg pizza in the oven for 10-15 minutes at 200°C till the base becomes golden brown and the cheese on top melts and gets browned.
- Serve vegetable pizza hot.

Tips for making veg pizza recipe

- The pizza dough can also be made in a food processor or electric mixer.
- If the dough becomes sticky add some flour and vice versa. If the dough is not soft and a bit dry, then add some warm water.
- A thin crust pizza with this dough can also be made and it was excellent.
- Can sauté the veggies lightly in olive oil if baker wants.
- The leftover pizza sauce can be refrigerated.
- Leftover pizza dough can be refrigerated or frozen. If frozen, it stays good for a month.

In text Questions

1. List the ingredients for the pizza preparation.

2. Enlist the steps are to be involved in pizza base.

3. Explain how to make vegetable pizza.

4. What are the points to be remembered while making pizza.

CHAPTER 18

PASTRIES

Types of Pastry

- Shortcrust Pastry. This is probably the most versatile type of pastry as it can be used for savoury and sweet pies, tarts and flans.

- Puff Pastry.

- Flaky Pastry.

- Rough Puff Pastry.

- Choux Pastry.

- Filo Pastry.

- Suet Crust Pastry.

- Hot Water Crust Pastry.

Pastry is a dough of flour, water and shortening (solid fats, including butter) that may be savory or sweetened. Sweetened pastries are often described as bakers' confectionery. The word "pastries" suggests many kinds of baked products made from ingredients such as flour, sugar, milk, butter, shortening, baking powder, and eggs. Small tarts and other sweet baked products are called pastries. Common pastry dishes include pies, tarts, quiches, croissants, and pastries.

Puff pastry, also known as pâte feuilletée, is a flaky, light pastry made from a laminated dough composed of dough (détrempe) and butter or other solid fat (beurrage). The butter is put in the dough (or vice versa), making a pat on which is repeatedly folded and rolled out before baking.

Principle of pastries

The basic principle to making pastry is a ratio of half fat (butter) to flour, with just enough water to bind it together. However, the more fat a baker can work into a baker, pastry the richer and flakier it becomes. 2. Ensure hands are not too warm – rinse under cold water if necessary.

Working Techniques of pastries

Pastry is so useful. The different types can be used in an endless array of dishes, savoury and sweet. There's a reason why there's such a thing as a specialist pastry chef, but don't be put off - it doesn't mean a baker can't make good pastry at home. Perhaps the most important thing is precision. This is no area for experimentation in the kitchen. Adapt a pastry recipe, and a baker are likely to find in a sticky situation. Weigh the ingredients carefully and use the best ingredients a baker can, especially flour.

A calm, organized approach using a cool worktop and ingredients will help produce good pastry. Work in as cool an environment as possible, well away from the cooker or boiler. Sift the flour, and when rubbing in the fat, lift the flour as high as possible while keeping it in the bowl - both steps will help incorporate cold air which will expand once in the oven and produce light pastry. Rub until the mixture looks like fine breadcrumbs, then stop - if a baker overdo it, a baker will have tough old pastry on hands. Add the water gradually; a baker really does not want to have to add more flour after the liquid. On the other hand, too little will create a dry and crumbly dough that will be hard to roll without breaking.

Leaving the pastry dough for half an hour or so to "relax" before rolling out (a baker can wrap it in cling film and place in the fridge) will give it greater elasticity and make it less likely to shrink when baking. When rolling out, work quickly and lightly, dusting the rolling pin to prevent sticking. Always roll away from a baker, turning the pastry, not the rolling pin. Rolling backwards and forwards, and side to side, will stretch the pastry and risk it shrinking in the oven. What's known as a "turn" is explained in the puff pastry recipe.

Pastry Making Faults

Table 3. Faults in pastry making

Subject	Key Information
Short Crust Pastry	Pastry is hard and has a tough texture:
	● Over kneading and heavy handling
	● Incorrect portions of ingredients
	● Too much water added
	Pastry is blistered:
	● Oven set temperature too high
	● Fat insufficiently mixed into flour
	● Uneven addition of water
	Pastry is fragile and crumbly:
	● Too much fat
	● Insufficient water
	● Overmixing the fat into flour
	● Pastry has sunk
	● Over working of pastry during kneading and rolling
Choux and Sue Pastry	Pastry has not flaked well:
	● Oven temperature too cool. Steam has not been produced.
	● Insufficient liquid added. Mixture was too dry
	● Pastry folded and rolled unevenly. Pastry has not rested sufficiently in a cool environment.
	● Pastry folded too thinly
	Shrinkage:
	● Dough not relaxed enough after rolling.

Short Crust Pastry

1. Sift 225g plain flour into a large bowl, add 100g diced butter and rub in with bakers fingertips until the mixture resembles fine breadcrumbs.

2. Stir in a pinch of salt, then add 2-3 tbsp water and mix to a firm dough.

3. Knead the dough briefly and gently on a floured surface.

4. Wrap in cling film and chill while preparing the filling.

Sweet Crust Pastry

Flour 100g

Fat 50 g (lard/butter/ margarine)

Castor sugar 50g

Egg to mix to a pastry roll out and shape/cut bake at 180°c

Take Flour 100 g. Rub in ¼ fat (lard/butter/ margarine), fold in the rest. Water to mix to a pastry. Roll out and shape/cut Bake. at 210°C. Flaky is much quicker to make. Puff requires a lot of resting. Folding develops the layers. A hot oven is needed to create steam to separate the layers. Strong flour with a lot of gluten (protein) can be used, to support the layers. Baking trays don't need oiling, If a baker mess up the rolling he can't 'screw' it back together as the baker will lose the layers quite a 'greasy' pastry due to fat content.

CHOUX PASTRY

Ingredients

Flour 75g

Fat 25g

Eggs 2

Water 125ml

- Made by a melting method – melt margarine/butter in water
- Bring to boil
- Add flour to make thick flour mixture

- Beat in eggs
- Bake in a hot oven – 210°c

Water must be boiling to allow the starch grains in the flour to Gelatinize (swell – burst- absorb liquid – thicken). Mixture must cool to below 60°c before adding eggs, otherwise they will scramble. Eggs must be beaten with a little at a time. Mixture should be stiff enough to pipe if needed. When out of oven stab the side with a sharp knife to allow steam to escape. Fill when cool. The classics are profiteroles and éclairs, but think about using different flavors for toppings and chopped fruit for healthier options. Use low fat yoghurt instead of cream to fill them. Savoury profiteroles are called 'aigrettes', any filling can be used, but think about lower fat options eg: cottage cheese, low fat mayonnaise.

SUET PASTRY

Ingredients

Self-rising flour 75g

Whole meal flour 50g

Suet (beef or vegetable) 75g

Milk 5-6 tbsp

Sugar (1.) 25g

Apples 4

Blackberries 100g

Sugar (2.) 3 tbsp

Water 3 tbsp (2)

Milk (to glaze) 5ml

●Important: Baker will need a medium sized pie dish to bake it in

Method

1. Peel and slice the apples.

2. Place apple slices in a pie dish with 3 tbsp of water and 3 tbsp sugar. Spoon over the blackberries and put to one side while baker makes the suet pastry.

3. Weigh the flours into a large (chilled) mixing bowl. Add the suet and stir with a palette knife.

4. Add the milk until a soft, soggy and sticky dough forms.

5. Now go in with baker's hands and bring it all together until baker have a smooth, elastic dough, which leaves the bowl clean.

6. Turn dough out onto a lightly floured surface.

7. Take pieces of pastry and roll into strips.

8. Dampen the rim of the pie dish with water, arrange the pastry strips over the fruit in a lattice pattern, pressing each end of the pastry strips onto the rim of the dish and trimming the ends neatly.

9. Brush the lattice with the 1 tbsp extra milk.

10. Brush the topping with milk to glaze.

11. Place on a baking tray and bake in oven at 190°C or Gas mark 5 for 25-30mins, until the topping is golden brown and the fruit is soft.

In text Questions

1. What is pastry?

2. Enlist the different types of pastries.

3. Explain how short and sweet pastries are prepared.

4. What is suet pastries?

5. Describe the faults in pastry making.

CHAPTER 19

PIES

A pie is a sweet or savory dish with a crust and a filling. The sides of a pie dish or pan are sloped. It can have a just a bottom, just a top, or both a bottom and a top crust. A tart is a sweet or savory dish with shallow sides and only a bottom crust.

While even tarts can be either sweet and savory, most of them are sweet. The filling in tarts is not covered. Tart crust is usually firm and crumble. Tarts tend to be thinner than pies. The various types of pie are: Pecan *Pie*, Apple *Pie*, Pumpkin *Pie*, Key Lime *Pie*, Cherry *Pie*, Lemon Meringue *Pie*, Sugar Cream *Pie*, Cheesecake.

Mixing Pie Dough

Prepare the Butter- Start with a recipe for the Flakiest Pie Crust. Cut butter into 1" pieces. Chill it while baker measure everything else.

Smash It Together- Mix the flour, sugar, and salt in a large bowl. Add the butter and toss until coated. Here's the fun part: Using baker's fingers and palms, work the butter into smaller, irregular pieces, moving quickly and aggressively so it stays cold. Flatten and thin, press some pieces and others that are larger and chunkier.

H_2O Break- Combine the vinegar and ice water in a measuring cup.

Hydrate the Dough- Drizzle the liquid over the flour mixture, running baker's fingers through the flour as baker goes to evenly distribute.

Knead in the bowl until dough starts to hold together. It will still look a little dry, but resist the urge to add more water; excess liquid can lead to a tough crust.

That's the look of a well-hydrated (but not overworked) dough

Knead the Dough- Turn the dough out onto a work surface (no extra flour needed) and smash with the heels of bakers hands a few more times, working on any shaggy edges. We still see largest pieces of butter and may be a dry spot here and there.

Stop Kneading- Over-working further develops the gluten, which will cause the crust to shrink when baked.

Cut, Pat, and Wrap- Cut dough in half. Press each half into a 1"-thick disk and wrap in plastic.

Chill- Chill at least 1 hour (and up to 3 days) to firm up the butter and allow the dough to hydrate, transforming this sorta-together lump into malleable pie dough. (Or freeze instead; it will keep up to a month.)

PECAN PIE

Deep dish pie shells (4-cup volume) 9" Cups

Karo corn syrup 1/ 2

Eggs 6

Slightly beaten 1 2 Cups sugar

Butter (or margarine) 4 Tbsp.

Melted 1/ 2 tsp. vanilla 1¼ 2½ Cups pecan halves

Preheat oven to 350°F.

Stir syrup.

Add eggs, sugar, butter and vanilla together, then stir in the pecans.

Pour into the pie shell(s) and bake for 50 to 55 minutes.

STRAWBERRY PIE

Pie shell 1 9"

Sugar 1 Cup

Water 1 Cup

Corn starch ¼ Cup

Strawberry Jell-O. 1 Box

Karo corn syrup 3 Tbsp.

Red food coloring ½ tsp.

Fresh strawberries 1 Container

Method

Cook sugar, water and corn starch until clear. Add ¼ cup Jell-O, syrup and food coloring. Cool. Cut strawberries; arrange on bottom of pie shell. Pour cooled mixture into shell. Put into fridge to set. When set, Garnish with cool whip and more berries.

PIE CRUSTS

Basic Ingredients

Homemade pie dough requires just a handful of ingredients — flour, butter (lard or shortening also work), salt, sugar and water — to achieve flaky goodness. The easiest way is to use a food processor. First, combine the dry ingredients. Then cut in the fat — make sure it's cold and cut into small pieces.

Add Water

Add ice water, 1 tablespoon at a time. Baker may not need the entire amount called for depending on the humidity. Pulse until it just comes together, then stop — don't over process.

151

Don't Overwork

If still see bits of butter, that is good. They will cook into buttery, flaky pockets of flavor.

Wrap It Up

Place the dough on a sheet of plastic wrap and fold into a package.

Ready to Roll

The last step before baking: Flatten the wrapped dough with a rolling pin and put it in the fridge for at least an hour, although overnight is best.

Common Problems in Fruit Pies

- Dough is too crumbly.
- Dough breaks when press it in the pan.
- Crust shrinks when it bakes.
- Crust is pale and under baked.
- Crust is too tough.
- (pie's) Bottom is soggy.
- Pecan pie's pecans have gone soft.
- Pumpkin pie is cracked

In text Questions

1. What is pie?
2. Write the procedure for pecan pie.
3. Explain on Pastry Making Faults.
4. Outline the procedure of pecan pie.

5. How do you process the Mixing Pie Dough ?

6. Point out the Common Problems in Fruit Pies.

7. List the Basic Ingredients of Pie Crusts. Explain.

CHAPTER 20

TARTS

Tarts are impressive desserts that look complicated and time-consuming, but baker can definitely replicate them at home. They are not as difficult to prepare as baker might think, as long as a baker streamline the process, which means baker should start with making the tart crust, let it chill while baker work on the pastry cream, let that chill while a baked roll out the dough, bake the tart in the oven while baker prepare the fruits, and finally, assemble all of those components. The greatest thing about this dessert is that the baker can make the tart crust and the pastry cream up to two days in advance, and assemble them right before the baker ready for serving.

Types

1. Chocolate Almond and Buckwheat Tart. This jaw-droppingly gorgeous creation is basically a fudgy brownie with a nutty buckwheat and almond crust.

2. Berries and Cream Tart.

3. Tomato Chevre Tart.

4. Strawberry Lemonade Tart.

5. Asparagus and Goat Cheese Galette.

6. Simple Blackberry Tart.

7. Raspberry Almond Crumb Tart.

8. Brown Butter Apple Tart.

French Fruit Tart with Vanilla Pastry Cream (makes one 9-inch tart or four 4 ½-inch tarts) Pâte brisée Crème pâtissière, Fresh fruits of choosing (such as berries, peaches, mangoes, apricots); ¼ cup apricot jam, seedless raspberry jam, or red currant jelly and 2 tbsp water.

Procedure for Making Small Fruit Tart

In a small saucepan, heat jam and water over low heat, stirring constantly, until it is melted. Remove from heat and set aside to cool. To assemble the tart, spread a layer of pastry cream evenly over the bottom of the tart crust. Arrange the fruits decoratively over the top and apply the glaze to the fruits with a pastry brush. Serve tarts immediately or cover and refrigerate for up to 12 hours. Tart is best enjoyed on the day it is assembled since the crust tends to get soggy overtime. The glaze is optional, but it does give the fruits an attractive sheen. Any dish that has a crust with a filling. There are four types of pies: cream, fruit, custard, and savory.

In text questions

1. What is tarts?
2. Write the different Types of Tarts.
3. What is the difference of pie and tart?
4. Bring out the Procedure for Making Small Fruit Tart.
5. How many types of pies? What are they?

CHAPTER 21

INTRODUCTION TO CONFECTIONERY

According to the Layman, Confectionery means "cakes and puddings, it does not limit itself to the mastery of skills of cake decoration but it is about so many things like different pastes, different chocolates, different types of sugar based products". Being fairly broad based, a closer look will reveal that it prepares the students not just for preparing few products, but also prepare them for a large segment of confectionery products.

CHOCOLATE

1. Growing Cocoa Beans

Chocolate begins with cocoa beans, the fruit of the cacao tree (also called a cocoa tree).

Scientists know that the cacao tree originated somewhere in South or Central America. Some say the first trees grew in the Amazon basin of Brazil, while others place its origin in the Orinoco Valley of Venezuela. Wherever its first home, we know the cacao tree is strictly a tropical plant thriving only in hot, rainy climates. Cocoa can only be cultivated within 20 degrees north or south of the equator.

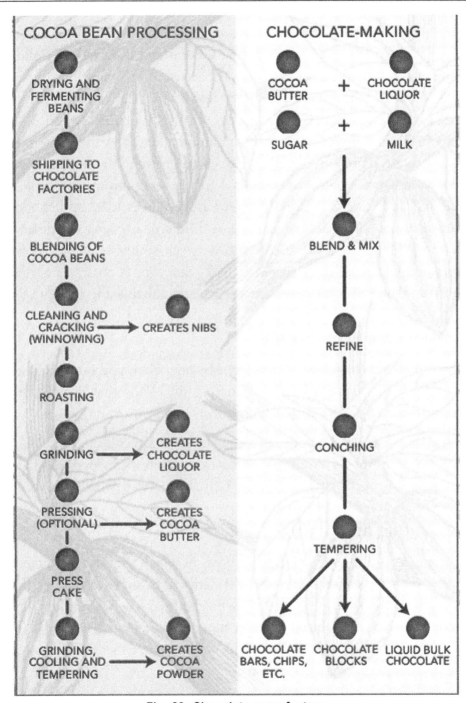

Fig. 29. Chocolate manufacture

Varieties of Cacao

There are two main species of cocoa: Criollo and Forastero. Criollo is sometimes called the prince of cacaos because it is a very high-quality grade of cocoa with exceptional flavor and aroma. Less than 15 percent of the world's cocoa is Criollo, grown mainly in Central America and the Caribbean. Forastero is a much more plentiful variety of high-quality cocoa, representing most of the cocoa grown in the world. Grown mainly in Brazil and Africa, it is hardier, more productive (higher yielding) and easier to cultivate than Criollo and is used in just about every blend of chocolate that is made. A third type of cocoa also deserves mention. Trinitario, a hybrid or cross between strains of the other two types, originated in Trinidad nearly 300 years ago. It possesses a good, aromatic flavor and the trees are particularly suitable for cultivation.

2. The Cocoa Bean harvest

Cocoa pods are harvested by hand, one by one. Each pod is carefully cut from the tree with a machete or sharp knife. Pods that grow on the tallest branches are harvested with knives attached to long poles.

After Picking, the pods cut from the trees are collected in piles in an open area not far from the cacao trees. Here the woody pods are opened with one or two lengthwise taps from a well-wielded machete.

Fermenting and Drying

Fermenting is a simple "yeasting" process in which the sugars contained in the beans are converted to acid, primarily lactic and acetic acids. The fermentation process takes from two-to-eight days, depending upon the cacao variety (Criollo beans ferment more quickly than Forestero). The beans are placed in large shallow wooden boxes or, on smaller farms, are left in piles and covered with banana leaves.

The drying process takes several days. Farmers or workers turn the beans frequently and use this opportunity to pick through them, removing foreign matter and flat, broken or germinated beans. During drying, beans lose nearly all their moisture and more than half their weight. When the beans are dried, they are ready to be shipped to chocolate factories around the world.

3. From Bean to Chocolate

The manufacturing process requires much time and painstaking care. Making an individual-size chocolate bar, for an instance, takes at least two-to-four days.

The pressed cocoa cake that remains after the cocoa butter is removed can be cooled, pulverized and sifted into cocoa powder. The powder is packaged for sale in grocery stores and in large quantities for commercial use as a flavor ingredient by dairies, bakeries and confectionery manufacturers.

While cocoa butter is removed to make cocoa powder, it must be added to make chocolate. This holds true of all eating chocolate, whether it is dark, bittersweet or milk chocolate. Besides enhancing flavor, the added cocoa butter makes the chocolate more fluid. One example of eating chocolate is sweet chocolate, a combination of unsweetened chocolate, sugar, cocoa butter and perhaps a little vanilla.

Whatever ingredients are used, the mixture then travels through a series of heavy rollers set one atop the other. These rollers press against the ingredients until the mixture is refined to a smooth paste ready for "conching."

Conching

Conching is a flavor development process which puts the chocolate through a "kneading" action. It takes its name from the conch shell-like shape of the containers originally used for this process. The "conches," as the machines are known, are equipped with heavy rollers that plow back and forth through the chocolate mass anywhere from a few hours to up to seven days. At this stage, flavorings are added if called for in the recipe. Conching develops the complex flavors and makes the chocolate velvety smooth.

After conching, the mixture is "tempered" — a process of carefully cooling the mixture while continually stirring it. Finally, the liquid chocolate is ready to be poured into molds shaped like the final product. The liquid chocolate also is used to enrobe (coat on all sides) certain chocolate bars such as those with whipped nougat centers and boxed chocolates which contain a variety of centers made from cream, fruit, nuts, and other ingredients.

CHOCOLATE BUTTER
Ingredients

½ cup unsalted butter, room temperature

½ cup semisweet chocolate chips, melted

1 tablespoon cocoa powder

salt

Method

Step 1: Beat butter, melted chocolate chips, cocoa powder, and a pinch salt with an electric mixer until smooth, about 1 minute.

Step 2: Serve at room temperature spread on toast or croissants.

WHITE CHOCOLATE
Ingredients

1/4 cup butter

1/3 cup icing sugar

2 tsp milk powder

1/4 tsp vanilla essence

1/8 tsp salt

Instructions

1. Boil water in a sauce pan, place a bowl over it. Make sure the bowl does not touch the boiling water. Now add butter.

2. Mix with a spoon until it melts.

3. Now add icing sugar little by little, keep mixing.

4. Mix and keep adding. Once baker is made with icing sugar, add milk powder.

5. Mix it well to avoid lumps. Add vanilla essence and salt.

6. Mix and switch off. See the consistency should be flowing. Now spoon the mixture into the chocolate mould.

7. Refrigerate at least for an hour. If baker touch and see it should not stick and should be firm.

8. Now demould the chocolates. Serve or store in fridge.

LIQUID CHOCOLATE

Ingredients:

- 300 ml fresh cream
- 300 gms whole chocolate
- 1 cup sterilized milk
- 100 gms cocoa powder

How to make Liquid Chocolate?

1. In a heating pan, add the fresh cream and stir continue until few boils.

2.   Now mix the chocolate and cocoa powder.

3.   Stir until they melt well in the cream.

4.   Pour a cup of milk and stir continue for 2 more minutes.

5.   Check the consistency before turning the gas off, as the consistency should be liquefying enough.

6.   Liquid Chocolate is ready to use.

Chef Tips

1.   Taste the liquid chocolate and baker may adjust with the sugar.

2.   If chocolate is not melting well then keep over the double boiler, in this way it will get melted properly.

3.   If liquid chocolate is too thick then add more amount of sterilized milk and then stir at the time of using.

4.   Liquid Chocolate can be poured on top of the vanilla ice cream to make it more delicious.

5.   Can use this liquid chocolate to make a brownie with ice cream recipe, just need to pour it hot.

FONDANT CHOCOLATES

50g melted butter for brushing

cocoa powder, for dusting

200g good-quality dark chocolate, chopped into small pieces

200g butter

200g golden caster sugar

4 eggs and 4 yolks

200g plain flour

Caramel sauce and vanilla ice cream or orange sorbet, to serve

Method

- First get moulds ready. Using upward strokes, heavily brush melted butter (use 50g in total) all over the inside of the pudding mould. Place the mould in the fridge or freezer. Brush more melted butter over the chilled butter, then add a good spoonful of cocoa powder into the mould. Dip the mould so the powder completely coats the butter. Tap any excess cocoa back into the jar, then repeat with the next mould.

- Place a bowl over a pan of barely simmering water, then slowly melt 200g good-quality dark chocolate and 200g butter, both chopped into small pieces, together. Remove the bowl from the heat and stir until smooth. Leave to cool for about 10 mins.

- In a separate bowl, whisk 4 eggs and 4 egg yolks together with 200g golden caster sugar until thick and pale and the whisk leaves a trail; use an electric whisk if baker wants. Sift 200g plain flour into the eggs, then beat together.

- Pour the melted chocolate into the egg mixture in thirds, beating well between each addition, until all the chocolate is added and the mixture is completely combined with a loose cake batter.

- Tip the fondant batter into a jug, then evenly divide between the moulds. The fondants can now be frozen for up to a month and cooked from frozen. Chill for at least 20 mins or up to the night before. To bake from frozen, simply carry on as stated, adding 5 mins more to the cooking time.

- Heat oven to 200°C.

- Place the fondants on a baking tray, then cook for 10-12 mins until the tops have formed a crust and they are starting to come away from the sides of their moulds. Remove from the oven, then leave to sit for 1 min before turning out.

- Loosen the fondants by moving the tops very gently so they come away from the sides, easing them out of the moulds. Tip each fondant slightly onto the bakers' hand so baker knows it has come away, then tip back into the mould ready to plate up.

- Starting from the middle of each plate, squeeze a spiral of caramel sauce – do all the plates baker need before baker goes on to the next stage.

- Sit a fondant in the middle of each plate. Using a large spoon dipped in hot water, scoop a 'quenelle' of ice cream.

- Carefully place the ice cream on top of the fondant, then serve immediately. Repeat with the rest of the fondants.

Toffee is made from sugar mixed with milk, butter or cream plus an ingredient such as lemon juice or golden syrup to stop it crystallizing. The mixture is heated to between 140^0C and 154^0C ('soft crack' stage and 'hard crack' stage), then allowed to cool and set.

Larger amounts of butter or cream can make chewier, softer toffees that are like caramels, while toffee cooked to a higher temperature becomes brittle but is still chewy to eat.

The Maillard reaction, which is caused by heating the dairy and sugar together, is what gives toffee its toasty flavor. The sugar has to be heated to 'soft crack' or 'hard crack' stage, which means baker will need to use a high-sided heavy-based pan and, ideally, a sugar thermometer. If baker do not have a thermometer, baker can test baker's toffee by dropping small amounts into iced water, then squeezing the ball that forms.

A ball of toffee that feels soft and squeezable will be at 'soft crack' stage. A firmer ball that's harder to shape will be at 'hard crack' stage. Some recipes cook the sugar at a lower temperature in order to make softer toffee.

There are a few key steps to follow:

1. Melt the sugar and butter together gently and evenly to avoid the butter separating out. Baker can stir the mixture while the sugar is dissolved – but once it has started to boil, stop stirring. Instead, tilt and swirl the pan.

2. Make sure bakers tin is prepared and sitting on a board or damp cloth before baker start. Once baker pours the hot toffee into the tin, it will heat up quickly.

3. Prepare all bakers ingredients in advance and have all baker's equipment to hand.

4. Take care at all times. Molten sugar will cause serious burns if it splashes on baker.

BUTTER TOFFEE RECIPE

Makes about 500g

- 300g golden caster sugar
- 300ml double cream
- 125g butter, cubed

1. Line the base and sides of a 20 cm × 30 cm baking tin with baking parchment and put it on a board.

2. Tip the sugar, cream and butter into a large, heavy-based, deep saucepan and heat gently, stirring occasionally, until all the ingredients have come together and the sugar and butter have melted.

3. Place a sugar thermometer or digital cooking thermometer in the pan, then turn up the heat and boil everything together vigorously, without stirring, until the temperature reaches 140^0C. Remove from the heat and leave for a moment to let any bubbles settle, then carefully pour the molten toffee into the prepared tin, swirling the

tin until the toffee fully covers the base. Leave for at least 2 hrs to set, or overnight if possible.

4. Use the baking parchment to lift the set toffee out of the tin, then cut the block into squares. If the toffee is sticking to the knife, lightly oil the blade. Wrap the toffee pieces in waxed paper. *Store in a jar for up to two weeks.*

BRITTLE TOFFEE RECIPE

Makes about 500g

- Oil, for the tin
- 450g golden caster sugar
- ¼ tsp cream of tartar
- 50g salted butter

1. Line the base and sides of an A4-sized tin with baking parchment, then oil it really well.

2. Put the sugar, cream of tartar, butter and 150ml hot water in a heavy-bottomed pan and heat gently until the sugar is dissolved, stirring occasionally.

3. Once the sugar has dissolved, turn up the heat and put the sugar thermometer in the pan.

4. Bring to the boil, then continue boiling until baker reach 'soft crack' stage on bakers thermometer (140^0C). This may take up to 30 minutes, so be patient. Don't leave the pan unattended as the temperature can change quickly. As soon as the mixture reaches 'soft crack' stage, tip it into a baker's tin and leave to cool.

5. Once cool, remove the toffee from the tin and break up with a toffee hammer or rolling pin. *Store in an airtight tin for up to a month.*

Storing toffee

Wrap toffees in wax paper or cellophane to stop them sticking to each other. Make sure they're stored in an airtight container in very dry conditions. Moisture will turn the surface of toffee, very sticky.

Toffee flavor

'English toffee' is a type of hard, buttery toffee popular in America that usually has a layer of chocolate and nuts on top. Cinder or honeycomb toffee (known as hokey pokey in New Zealand) has baking soda and vinegar added so the toffee froths and makes bubbles as it sets, giving a completely different texture.

Fig. 30. Nutty toffee

Caramel, which is melted, caramelized sugar with nothing added, is sometimes referred to as toffee – for instance, when used in toffee apples or in bonfire toffee recipe. When mixed with nuts, it's often known as brittle.

Although the flavor is popular in all sorts of dishes, lots of 'toffee' recipes don't actually contain toffee; sticky toffee pudding, for example, has a toffee-flavored sauce but doesn't require baker to make toffee first. Shards of toffee work well in cakes, biscuits and other bakes and can be stirred into ice cream. Toffee can also have flavor added, from salt (salted caramel) to alcohols (such as Baileys) and spices, nuts and dried fruit.

Additionally, cocoa contains stimulant substances like caffeine and theobromine, which may be a key reason why it can improve brain function in the short term. Cocoa or dark chocolate may improve brain function by increasing blood flow.

Baking chocolate

Baking chocolate (also known as unsweetened chocolate or bitter chocolate) comes in a bar, but it is no candy bar: There's no sugar in this and it is super bitter. Basically, it's the essence of chocolate: solidified 100-percent chocolate liquor (the center of cocoa beans, ground to a liquid) without added sweeteners, flavors and emulsifiers. Baking chocolate comes in many varieties, including semi-sweet and milk. But unless bakers recipe says otherwise, use unsweetened in brownies, cakes, frosting and the like, adding sugar separately. This gives baker the most control over the sweetness. Conversely, don't use baking chocolate in recipes that don't mix it with sugar, like candy coating. And don't chop it up for chocolate-chip cookies. Just. Don't.

Dark Chocolate (Includes Both Semisweet and Bittersweet)

Dark chocolate is chocolate liquor that's been fancied up with extra cocoa butter, sugar, emulsifiers and flavorings. It retains a high percentage of cacao — anywhere from 65 to 99 percent. The higher the percentage, the less sweet. There are several kinds of dark chocolate, all with different ratios of sugar to cocoa. None contain milk solids, which is excellent news for vegans.

Semisweet chocolate and bittersweet chocolate are types of dark chocolate that contain at least 35 percent chocolate liquor. Bittersweet usually contains more cacao than semisweet, which is sweeter.

Dark chocolate can be eaten straight up or used in recipes for garnishes, icings, glazes and cookies. Semisweet should be bakers default for chocolate-chip cookies. Some dark chocolate can be quite expensive.

Milk Chocolate

Just like the name suggests, milk chocolate does contain dairy. It's commonly made by adding dry milk solids (like powdered milk) to the chocolate. At around 55 percent sugar and 20 percent cocoa butter, this creamy variety of chocolate is mild and quite sweet.

Milk chocolate melts easily. Baker certainly can use it in doughs and batters, but it's easiest to handle in no-bake recipes such as sauces, fillings or icings, or as a topping for already-baked treats.

One word of warning: Milk chocolate's high sugar content makes it sensitive to heat, so it may burn if baker tries to use it in recipes that call for semisweet chocolate.

White Chocolate

White chocolate is made of sugar, milk and cocoa butter, but without the cocoa solids. Its ratios are actually quite close to milk chocolate's, but the absence of cocoa solids gives it a creamy, ivory hue. White chocolate's sweetness makes it a great addition to baked goods, which typically call for less sugar to compensate. But don't substitute it in for dark or baking chocolate, as it may burn. It's also pleasing as a candy coating or in icings and ganache.

Chocolate Morsels

These are the little "chocolate kiss"-shaped morsels that baker stir into cookies — sold in dark, milk and white.

Morsels will not melt entirely; they are meant to hold their chip-like shape. So fold them into the batter or dough, or use as a topping.

Chocolate-Flavored Coating

Coating chips use vegetable fats to supplement (or replace) the cocoa butter. Technically, they're not really chocolate at all. They do have a slight chocolate flavor — though sometimes they can also taste waxy.

Chocolate coating melts well, so it's useful for making truffles, cake pops or other treats. It's also rather malleable, so it works well in shape molds.

Cocoa Powder (Includes Dutch Process Cocoa)

Cocoa powder is made from ground cocoa solids that don't contain any cocoa butter. There are two key types: regular and Dutch process. The regular variety is sold as sweetened (for hot cocoa and such) and unsweetened (the kind more frequently used in baking). This cocoa reacts with alkali ingredients such as baking soda, helping give baked goods a lift.

Fig. 31. Dark chocolate

Dutch process cocoa (also sold as "dutched" or "alkalized") has been treated with an alkaline solution to neutralize acidity. This process darkens the color and makes the flavor milder.

In text questions

1. How to make toffee safely?

2. How to Make Eating Chocolate?

3. What type of chocolate helps bakers body?

4. What is Conching?

5. How milk and White Chocolate are made?

6. What is Chocolate Morsels?

CHAPTER 22

PREVENTION OF BACTERIAL ROPE AND MOLD INFECTION

The growth of microbes slows down such as yeast and molds when the bread is stored in a clean, dry container that allows the free flow of air. The air will stop evaporating moisture from inside the bread making the surface damp and encouraging microbial growth. Bread should be eaten within a few days.

The most common source of microbial spoilage of bread is mold growth. Less common, but still causing problems in warm weather, is the bacterial spoilage condition known as 'rope' caused by growth of Bacillus species. Mold spoilage of bread is due to post-processing contamination.

The most troublesome bacterial contaminants during fermentation are members of the lactic acid bacteria, such as Lactobacillus and Pediococcus, which cause diacetyl formation, lactic acid formation (Lactobacilli) and ropiness (Pediococci).

Foods that are potentially hazardous inside the danger zone:

- Meat: beef, poultry, pork, seafood.
- Eggs and other protein-rich foods.
- Dairy products.
- Cut or peeled fresh produce.

- Cooked vegetables, beans, rice, pasta.

- Sauces, such as gravy.

- Sprouts.

- Any foods containing the above, e.g. casseroles, salads, quiches.

- Hot holding means keeping the food in any equipment that is designed to keep the food hot. It is important to make sure that the food is safe to eat by maintaining proper temperature. The main function of a hot holding unit is to maintain the temperature of each food at 140°F or warmer.

A Moist Place- Moisture spreads the yeast through bread and also is necessary for mold to grow. Bread is a fairly dry food, but there is always some moisture in it, because water is how the "good" mold (yeast) spreads throughout the loaf and makes it rise.

Spoilage bacteria can cause fruits and vegetables to get mushy or slimy, or meat to develop a bad odor, but they do not generally make baker sick. Pathogenic bacteria cause illness

In addition to yeast, bacteria play an important role in sourdough bread making. Starches in flour, when water is added, break down into sugars with the help of enzymes (which are molecules and not organisms). The yeast feed on some of the sugars and make carbon dioxide that raises the dough.

Method

1. Invest in a breadbox. Place baker's breadbox in a cool area away from heating elements that can accelerate mold growth.

2. Keep the bread dry. Avoid touching the loaf with wet hands and never seal the loaf with visible moisture around it.

3. Avoid the refrigerator. Storing in Cool, Dry Place

Bread and other bakery products are subjected to various spoilage problems, viz., physical, chemical and microbial; the latter is the most serious one particularly bacterial (Bacillus sp.) and mold growth.

For food safety experts, moldy bread is bad. Some molds, like those used for Gorgonzola cheese, are safe to eat. But the mold dotting bread is not a benign source of extra fiber. Gravely says people who eat moldy food may suffer allergic reactions and respiratory problems.

Although baker might think all microbes are nasty and cause diseases like colds and flu, most microbes are actually harmless and some are even useful. Making bread is a brilliant example of this. Tiny fungi, called yeast, are added to bread dough. The bread dough is made of flour and water.

Ropiness is bacterial spoilage of bread that initially occurs as an unpleasant fruity odor, followed by enzymatic degradation of the crumb that becomes soft and sticky because of the production of extracellular slimy polysaccharides

By keeping the bread in a cool and dark place, it will last longer and stay fresh. Heat, humidity and light are all bad for bread, but great for fungi or mold, so consider bakers' fridge is the best to keep bread fresh and delicious. Tightly sealing the bread also helps to slow the molding process.

In text questions

1. How does baker prevent microorganisms from growing on the bread?

2. What are the spoilage organisms seen in baked food products? What causes bread spoilage?

3. Which bacteria causes Ropiness in bread?

4. What is the main function of a hot holding unit?

5. What causes bread to mold?

6. Can rope spoilage make baker sick?

7. How are bacteria used to make bread?

8. How does baker preserve bread from fungal spoilage?

9. Does bread grow bacteria?

10. Can baker eat moldy bread?

11. What microbes are in bread?

12. What is bread Ropiness?

13. What prevents mold from growing on bread? Explain types of sugar

CHAPTER 23

EQUIPMENT USED IN BAKERY AND CONFECTIONERY

Small Equipment used in Bakery and Confectionery

Measuring Jug: An equipment used for measuring all the types of liquids in the liter.

Biscuit Cutter: It is used for the cutting of different types of biscuits. These are available in different fancy shapes.

Wooden Spoon: It is used at the time of cooking, especially sugar based products.

Wire Whisker: It is used for whisking egg and cream and helps to aerate with air.

Turn Table: It is used while icing on the cakes and pastries.

Scrapper: It is used while creaming and dough making to collect the raw material.

Pizza Cutter: It is a cutter used for cutting the pizza and sometimes to cut the rolled dough.

Doughnut Cutter: It is used for cutting the rolled doughnut dough.

Icing Comb: It is used while doing the cream icing on the cakes.

Rolling Pin: It can be of different material and of different lengths, used for rolling the dough.

Nozzle Set: It is used for the decorative work on cakes, cookies and different products.

Strainer: It is used for straining the liquids to remove impurities.

Spatula: It can be of wooden, plastic or rubber material and is used for removing batter or mixture from the machine bowl.

Piping Bag: It is used while piping the batters, cookies mix, cream icing etc.

Basin: A large bowl used for making of dough, batter or storage of food.

Bread Mold: A mold used for preparing the molded breads.

Tart Mold: A mold used for the preparation of tarts.

Muffin Tray: A kind of baking tray for baking the batter of muffins.

Caramel Custard Mold: A mold used for the making of Caramel custard.

Fancy Mould: It is used for the preparation of different fancy cakes.

Cake Mould: It is used for baking the cake batter.

Flan Mould: An equipment used for the making of flans.

Laddle: An equipment used for the portioning of raw material and also for cooking.

Pallet Knife: A knife with parallel and without any sharp edges. Used for the different products like cakes, icing etc.

Bread Knife: A long knife with one edge with the grooved like saw, used for cutting of cakes and breads.

Measuring Spoon: It is used for measuring the dry ingredients in small quantity like 1.5 gms, 2.5 gms, 5 gms, 10 gms.

Baking Tray: It is used for the different baking like-breads, biscuits, pizza etc.

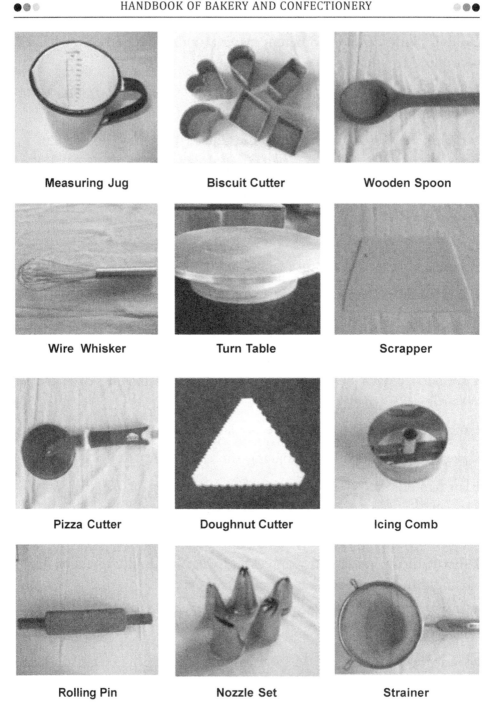

Measuring Jug	Biscuit Cutter	Wooden Spoon
Wire Whisker	Turn Table	Scrapper
Pizza Cutter	Doughnut Cutter	Icing Comb
Rolling Pin	Nozzle Set	Strainer

Fig. 32A. Small equipment used in bakery and confectionery

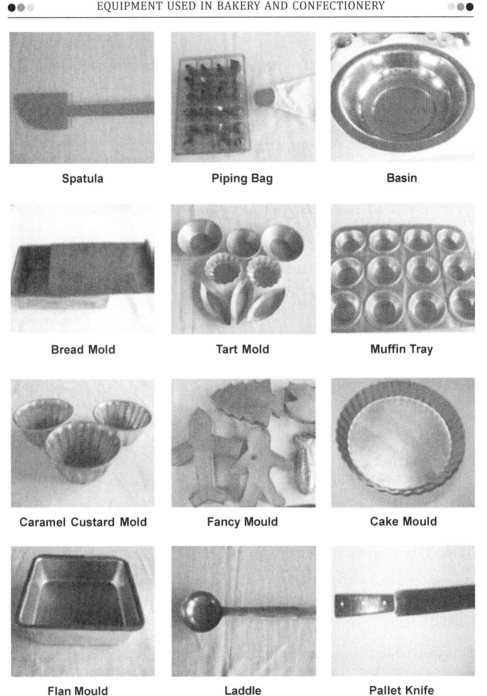

Fig. 32B. Small equipment used in bakery and confectionery

Bread Knife   **Measuring Spoon**   **Baking Tray**

Fig. 32C. Small equipment used in bakery and confectionery

Large Equipment used in Bakery and Confectionery

Weighing Scale: It is used for the weighing the raw materials in the unit of grams and kilograms.

Single Deck Oven: It is an oven with the single deck used for baking.

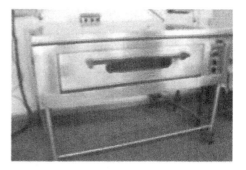

Weighing Scale   **Single Deck Oven**

Fig. 33A. Large equipment

Table Top Planetary Mixer

An equipment with the three attachments - kneader, whisker and creamer for different methods of preparations in bakery and Confectionery.

Dough Divider: An equipment used for dividing the dough into equal weights.

Two Deck Oven: It can be used for baking two different products at different baking temperatures.

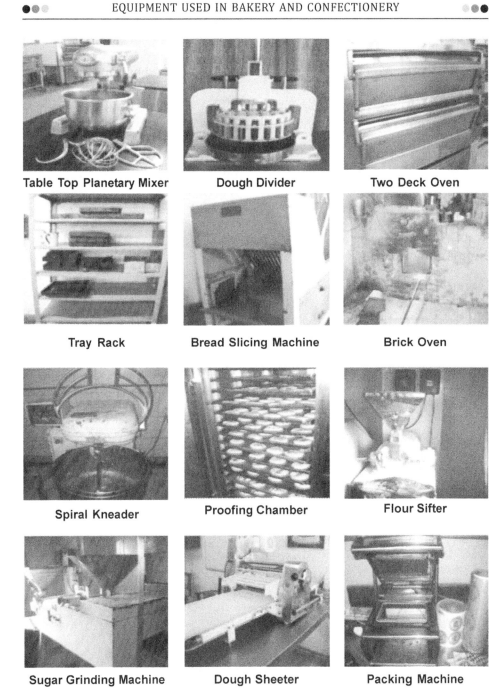

Table Top Planetary Mixer

Dough Divider

Two Deck Oven

Tray Rack

Bread Slicing Machine

Brick Oven

Spiral Kneader

Proofing Chamber

Flour Sifter

Sugar Grinding Machine

Dough Sheeter

Packing Machine

Fig. 33B. Large equipment

Tray Rack: A rack to place the baked products and baking trays.

Bread Slicing Machine: A machine used for the slicing of the bread and cake loafs.

Brick Oven: An old style oven made of bricks, where wood and charcoal to be used for heating the oven.

Spiral Kneader: A kneader used for the bulk kneading.

Proofing Chamber: A cabinet used for proofing the dough, having a humidity controller.

Flour Sifter: An equipment used for the shifting flour in bulk quantity.

Sugar Grinding Machine: A sugar grinder for bulk grinding of sugar.

Dough Sheeter: An equipment used for the sheeting of dough to a desired thickness.

Packing Machine: An equipment used for the packing of prepared products which is for the sale.

SUGGESTED READINGS

- Bushuk W, Rasper V. Wheat: Production, Composition and Utilization. Blackie Acad and Professional, Glasgow. 1994.

- D' Appolonia BL, Kunerth WH. The Farinograph Handbook, AACC, St Paul, MN, USA 1984.

- Dobraszczyk BJ, Dendy DAV. Cereal and Cereal Products: Chemistry and Technology. Aspen Publisher, Inc Maryland. 2001

- Eliasson, A.C. and Larsson, K. Cereals in Bread making, Marcel Dekker, Inc. New York. 1993.

- Fance,Wilfred. J., ed. The New International Confectioner, 5th ed. London:Virtue & Co.,1981.

- Finney K, Yamazaki WT . Wheats and Wheat Improvement. American Society of Agronomy. Madison WI. 1967.

- France.W.J: The student Technology of Bread making and flour confectionery, Routledge and Kegan Paul Ltd., London, UK. 1974.

- Hammer RJ, Hosney RC . Interactions: the key to cereal quality. AACC, St Paul, MN 1998.

- Heyne EG . Wheat and wheat improvement. American Society of Agronomy. Haworth Press Inc, Canada. 1987.

- Hoseney, R.C. Principles of Cereal Science and Technology, Am. Assoc. Cereal Chemists, St. Paul, MN, USA. 1986.

- Hudson, Biochemistry of Food Proteins, 2002.

- John Kingslee, A professional text to Bakery and Confectionery, New Age International (P) Limited. 2014.

- Kasarda DD, Bernardin JE, Nimmo CC. Advances in cereal science and technology. Vol 1. AACC, St Paul, MN 1976.

- Kent.N.L.: Technology of cereals – with special reference to wheat, pergamon Press, New York, USA. 1975.

- Malik. R.K. and Dhingra. K.C.: Technology of Bakery Industries. Small Industry Research Institute, New Delhi, India. 1981.

- Manay Shakunthala, N and Shadaksharaswamy M. Food Facts and Principles, New Age International (P) Ltd Publishers, Reprint 2005.

- Matz S.A.: Bakery Technology, packaging, nutrition, product development and quality assurance, Elsevier Science Publisher Ltd., New York, USA. 1989.

- Matz. S.A. Technology for the Materials of Baking, Elsevier Science Publishers. Baking, England. 1989.

- Neelam Khetarpaul, Raj Bala Grewal and Sudesh Jood, Bakery science and cereal technology, Daya publishing house. 2013.

- NIIR Board of consultants and engineers, The complete technology book on bakery products, second edition, National Institute of Industrial Research, Delhi. 2009.

- Norman, N.P and Joseph, H.H. Food Science, Fifth edition, CBS Publication, New Delhi. 1997.

- Phillips RD and JW Finley. Protein Quality and the effects of processing, Marcel Dekker, New York. 1989.

- Pomeranz, Y. Advances in Cereal Science and Technology, Am. Assoc. Cereal Chemists, St. Paul, MN, USA .1976.

- Pomeranz, Y. Wheat is Unique. AACC Inc. St. Paul MN. USA 1989.

- Pomeranz, Y. Wheat: Chemistry and Technology, Vol. I, 3rd Ed., Am. Assoc. Cereal Chemists, St. Paul, MN, USA. 1988

- PomeraNz, Y.: Wheat Chemistry and Technology, Vol. 1 and II American Assn. of Cereal Chemists, 3rd Ed. St. Paul Minnesota, USA. 1998.

- Salunkhe, D. and Despande, S.S. Foods of Plant origin: Production, Technology & Human Nutrition, An AVI Publications, New York. 2001.

- Samuel, A.M. The Chemistry and Technology of Cereal as Food and Feed, CBS Publishers & Distribution, New Delhi. 1996.

- Sivasankar, B. Food Processing and Preservation, Prentice Hall of India Pvt. Ltd, New Delhi. 2002.

- Sultan.W: Practical baking manual – for students and instructors, AVI Publishing Co.INC, West Port, Connecticut. 1976.

- Vijaya Khader, Text book of Food Science and Technology, Indian Council of Agricultural Research, New Delhi, 2001.

- Yogambal and Ashok kumar, Theory of Bakery and Confectionery, PHT Learning Private Limited, New Delhi. 2009.